Deep-Space Probes

Springer
*London
Berlin
Heidelberg
New York
Barcelona
Hong Kong
Milan
Paris
Santa Clara
Singapore
Tokyo*

Gregory L. Matloff

Deep-Space Probes

 Springer

Published in association with
Praxis Publishing
Chichester, UK

Dr Gregory L. Matloff
New York University
General Studies Program
New York
USA

SPRINGER–PRAXIS BOOKS IN ASTRONOMY AND SPACE SCIENCES
SUBJECT *ADVISORY EDITOR*: John Mason B.Sc., Ph.D.

ISBN 1-85233-200-X Springer-Verlag Berlin Heidelberg New York

British Library Cataloguing in Publication Data
  Matloff, Gregory L.
    Deep-space probes
    1. Space probes 2. Outer space – Exploration
    I. Title
    629.4'354
    ISBN 1-85233-200-X

Library of Congress Cataloging-in-Publication Data
  Matloff, Gregory L.
    Deep-space probes/Gregory L. Matloff.
      p. cm. – (Springer–Praxis books in astronomy and space sciences)
    Includes index.
    ISBN 1-85233-200-X (alk. paper)
      1. Space probes. I. Title. II. Series.
  TL795.3.M38 2000
  629.43'5–dc21   00-061255

Apart from any fair dealing for the purposes of research or private study, or criticism or review, as permitted under the Copyright, Designs and Patents Act 1988, this publication may only be reproduced, stored or transmitted, in any form or by any means, with the prior permission in writing of the publishers, or in the case of reprographic reproduction in accordance with the terms of licences issued by the Copyright Licensing Agency. Enquiries concerning reproduction outside those terms should be sent to the publishers.

© Praxis Publishing Ltd, Chichester, UK, 2000
Printed by MPG Books Ltd, Bodmin, Cornwall, UK

The use of general descriptive names, registered names, trademarks, etc. in this publication does not imply, even in the absence of a specific statement, that such names are exempt from the relevant protective laws and regulations and therefore free for general use.

Copy editing and graphics processing: R. A. Marriott
Cover design: Jim Wilkie
Typesetting: Originator, Great Yarmouth, Norfolk, UK

Printed on acid-free paper supplied by Precision Publishing Papers Ltd, UK

This book is dedicated to all those visionaries and dreamers who have dreamed of flight beyond the confines of Earth's surface and travel to the stars. From the myth of Daedalus and Icarus, the early aircraft designs of Leonardo, the inspiring science fiction of many masters of the craft, to the originators of modern-day astronautics, these men and women have established the philosophical and technological foundations upon which the first starships will be designed. I fervently hope that some of the spirit of these pioneers has entered into this book.

Table of contents

Foreword	xi
Author's preface	xiii
Acknowledgements	xv
List of illustrations	xvii
Introduction	xxi

1 MOTIVATIONS FOR DEEP-SPACE TRAVEL 1
   1.1   An inventory of near-Earth objects 4
   1.2   Considering NEO origins 6
   1.3   The difficulty of telescopic exploration of NEOs near their point of origin 7
   1.4   Robotic exploration options 9
   1.5   Directed panspermia 9
   1.6   Bibliography 10

2 THE REALMS OF SPACE
                                                                                  13
   2.1   The question of origins 13
   2.2   Realms of fire, water and ice 15
   2.3   Solar radiant flux and planet effective temperature 16
   2.4   The effect of atmospheric optical depth 18
   2.5   The lifetime of a planet's atmosphere 18
   2.6   Comparative planetology: an appreciation of the lifezone 20
   2.7   Beyond the planets: the reefs of space 21
   2.8   Bibliography 22

3 TOMORROW'S TARGETS   25
   3.1   The TAU mission: an early NASA/JPL extrasolar mission study   26
   3.2   SETIsail and ASTROsail: proposed probes to the Sun's gravity focus 27

	3.3	The Aurora project: a sail to the heliopause	29
	3.4	The NASA Interstellar Initiative	30
	3.5	The NASA heliopause sail	30
	3.6	The NASA Kuiper Belt explorer	31
	3.7	A probe to the Oort Cloud	32
	3.8	Bibliography	33
4	**SPACE PROPULSION TODAY**		**35**
	4.1	Rocket history and theory	35
	4.2	The solar-electric drive	39
	4.3	Unpowered planetary gravity assists	40
	4.4	Powered solar gravity assists	44
	4.5	The solar-photon sail	46
	4.6	Bibliography	54
5	**THE INCREDIBLE SHRINKING SPACEPROBE**		**57**
	5.1	The small and the very small	58
	5.2	Nanotechnology: the art and science of the very, very small	59
	5.3	Nanotechnology and spaceflight: near-term possibilities	61
	5.4	Nanotechnology and spaceflight: long-term possibilities	62
	5.5	Possible limits to nanotechnology	63
	5.6	Bibliography	64
6	**THE NUCLEAR OPTION**		**65**
	6.1	Nuclear basics	65
	6.2	Nuclear-electric propulsion (NEP)	69
	6.3	Nuclear-pulse propulsion: Orion, Daedalus and Medusa	72
	6.4	Inertial electrostatic confinement and gas-dynamic mirror fusion	75
	6.5	Antimatter: the ultimate fuel	76
	6.6	Bibliography	79
7	**TWENTY-FIRST CENTURY STARFLIGHT**		**83**
	7.1	Laser/maser sailing fundamentals	84
	7.2	Starship deceleration using the magsail	87
	7.3	Thrustless turning	90
	7.4	Perforated light sail optical theory	92
	7.5	The Fresnel lens: a method of improving laser beam collimation	94
	7.6	Round-trip interstellar voyages using beamed-laser propulsion	96
	7.7	Interstellar particle-beam propulsion	97
	7.8	Bibliography	98
8	**ON THE TECHNOLOGICAL HORIZON**		**101**
	8.1	The hydrogen-fusing interstellar ramjet	101
	8.2	The ram-augmented interstellar rocket (RAIR)	104
	8.3	The laser ramjet	108

	8.4	The ramjet runway	110
	8.5	A toroidal ramscoop	112
	8.6	Bibliography	115

9 EXOTIC POSSIBILITIES ... 117
9.1 'Shoes and ships': the potential of magnetic surfing ... 118
9.2 'Sealing wax': approaches to antigravity ... 122
9.3 'A boiling-hot sea': zero-point energy and special-relativistic star drives ... 122
9.4 'Cabbages and kings': general relativity and spacetime warps ... 126
9.5 'Winged pigs': some other exotic ideas ... 128
9.6 Bibliography ... 128

10 OF STARS, PLANETS AND LIFE ... 133
10.1 A short history of extrasolar planet detection efforts ... 133
10.2 Methods of imaging extrasolar planets ... 135
10.3 Extrasolar planets found to date that orbit Sun-like stars ... 138
10.4 How common are life-bearing worlds? ... 140
10.5 Terraforming: a way to increase the odds ... 142
10.6 Bibliography ... 143

11 LIFE BETWEEN THE STARS ... 145
11.1 Environmental obstacles to interstellar flight, and their removal ... 146
11.2 Options for onboard power between the stars ... 149
11.3 Close environment life support systems ... 150
11.4 Of worldships and interstellar arks ... 151
11.5 Hibernation for humans: the long sleep to α Centauri ... 153
11.6 Bibliography ... 154

12 CONSCIOUS SPACECRAFT ... 157
12.1 The von Neumann machine: can the computer eat the Galaxy? ... 158
12.2 The cryogenic star-child ... 159
12.3 The virtual star-traveller ... 159
12.4 Bibliography ... 160

13 MEETING ET ... 161
13.1 Are starships detectable? ... 162
13.2 Motivations of star-travelling extraterrestrials ... 164
13.3 Bibliography ... 165

Afterword ... 167
Nomenclature ... 169
Glossary ... 175
Index ... 178

Foreword

This excellent book by Dr Gregory Matloff could be viewed as a large multi-disciplinary compendium of past research, current investigations and future research in astronautics. However, unlike conventional works that are usually closed, this book is an *open* guide in three main respects: it contains progressive exercises as the chapter and section topics evolve, it provides the reader with many updated references, and it clearly indicates projected research areas that could become *current* research in the near future.

For a student, the included exercises could be transformed into small worksheets or notebooks featuring many modern symbolic or algebraic computation systems that run on desktop or laptop computers. Thus, rapid and progressive study is possible – a sort of learning library driven by the author and by the bibliography at the end of each chapter. More than 340 references for both professionals and students have been selected to provide the reader with a sound basis for expanding his or her knowledge of the many different subjects dealt with authoritatively by the author, who for several decades has worked actively in various fields of astronautics. A systematic and unhurried study of this book will produce for the reader an extraordinarily enlarged number-based view of spaceflight and its significant impact on our future global society.

If conscious life is to eventually expand slowly and inexorably, and without limit, into space, it is important that appropriate scientific mentalities expand in advance in to human brains. This book is a positive contribution at the beginning of such an exciting though complex read.

Giovanni Vulpetti
International Academy of Astronautics
1 September 2000

Author's preface

In light of a recent paradigm shift at NASA, there can be little doubt that humans or their robot proxies are heading towards the far reaches of the Solar System and the Galaxy beyond. A number of books on the subject of interstellar flight are currently on the market (one of them co-authored by myself).

But this book is designed to be unique in at least several respects. First, I have attempted to be as up-to-date as possible, drawing upon my experience as a Faculty Fellow of the NASA Marshall Spaceflight Center in Huntsville, Alabama, where I contributed to NASA interstellar research during the summers of 1999 and 2000. Research results presented at recent international conferences in this field are reviewed, as well as articles in the peer-reviewed literature.

A few years ago it was possible for an individual to be acquainted with all recent work in interstellar studies; but the field has recently expanded explosively and such comprehensive insight is no longer possible, and I apologise if I have omitted your favourite paper or propulsion system.

Because one component of this book's potential audience is the experienced astronautical or astronomical professional, I have not hesitated to include higher mathematics in considering the performance of various deep-space propulsion techniques and observational methods. Many exercises dealing with the equations are also included. It is hoped that engineering or physics professors might use these as teaching aids, and undergraduate physical science or engineering students interested in participating in deep-space exploration might also investigate them. Before you can invent the ultimate star drive, you might find value in gaining mathematical facility with previously suggested ideas.

As many others have suggested, our continued evolution as a species may require a holistic recognition that we are of the same material as the stars, and our expansion into space may be an unconscious reaction to this – starstuff returning to the stars. Quite possibly we must leave the Earth in large numbers if we are to perserve her. Are we in some ways returning home, by going back to the origins of our past, before the Earth was formed? If we are part stardust, part terrestrial, are we meant to spread the seeds of Gaia to other planets and other star-systems? In order to do this we must function in the more connected way of a focused organism seeking it's destiny, recognising that going outward is in some way our return? A global effort will be needed to take on this enormous technological and spiritual challenge, and we may well abandon our little political, religious, cultural selves in order to claim our larger destiny.

Acknowledgements

If you find this work enjoyable, beautiful and informative, many people deserve credit. In particular, I would like to thank Brice Cassenti and Giovanni Vulpetti, who expertly read the draft manuscript, copy editor Bob Marriott, and Clive Horwood, Chairman of Praxis. Without their efforts, publication of this volume would have been impossible.

Many other individuals have participated in the author's education in various aspects of astronautics, physics and astronomy. A partial list includes Buzz Aldrin, Edwin Belbruno, Alan Bond, John Cole, Robert Forward, Giancarlo Genta, Martin Hoffert, Anders Hansson, Les Johnson, Eugene Mallove, Claudio Maccone, Vincenzo Millucci, Seth Potter, Carl Sagan, Salvatore Santoli, Peter Schenkel, George Schmidt, Gerald Smith, Jill Tarter, Arthur Upgren and Jeffrey Van Cleve. These people, and other cited experts, deserve much of the credit for what is good in this book. The author bares full responsibility for any errors.

The poetry and prose quotes at the beginning of each chapter have been selected to add a visionary quality to the text. The quote used in the introduction is from the Dover Publications 1968 edition of Olaf Stapledon's *Last and First Men* and *Star Maker*, and is reproduced by permission of the publisher.

Loren Eisley's *The Invisible Pyramid* (Scribner's, New York, 1970) is the source for the quote at the beginning of Chapter 1. I am grateful to Simon & Schuster in New York for granting me permission to quote from this book.

The Chapter 2 quote is from *Planet Quest: The Epic Discovery of Alien Solar Systems*, by Ken Croswell (copyright © 1997 by Ken Croswell). This is reprinted with permission of The Free Press, a division of Simon & Schuster.

Arthur C. Clarke's classic *The Promise of Space* (Harper & Row, New York, (1968) is the source for the quote that heads Chapter 6. Permission to quote was granted by the Scovil, Chichak & Galen Literary Agency in New York. The quotes that introduce Chapters 7 and 13 are from another classic – Carl Sagan's *Cosmic Connection*, (Anchor/Doubleday, Garden City, NY, 1973). This book, which was produced by Jerome Agel, was subsequently retitled *Carl Sagan's Cosmic*

Connection, by Carl Sagan, produced by Jerome Agel (copyright © Carl Sagan and Jerome Agel). Permission to quote was granted by Jerome Agel of New York.

What better way to start Chapter 8 on the interstellar ramjet than a quote from Poul Anderson's novel *Tau Zero* (Doubleday, New York, 1970)? Permission to quote was granted by Random House Inc., New York. Permission for the Chapter 10 quote from Robery Ardrey's *African Genesis* was granted by the late author's son Daniel Ardrey.

The Chapter 12 quote is from *2001: A Space Odyssey* by Arthur C. Clarke, (copyright © 1968 by Arthur C. Clarke and Polaris Productions Inc.) Permission to quote was granted by Dutton Signet, a division of Penguin Putnam, Inc., New York.

List of illustrations

Chapter 1
1.1   Cumulative percentage of known near-Earth objects   5
1.2   Geometry for remote viewing of a small celestial object.   8

Chapter 2
2.1   Factors influencing a planet's habitability   20

Chapter 3
3.1   The Sun as a gravitational lens.   27
3.2   An inflatable solar-sail probe to the Sun's gravity focus.   28

Chapter 4
4.1   The rocket principle   37
4.2   The solar-electric drive   39
4.3   Representation of an unpowered planetary flyby.   41
4.4   The geometry of a flyby of a celestial object   42
4.5   Variations of the solar-photon sail.   47
4.6   Gradual unfurlment of an inward-bent solar sail as it recedes from the Sun   48
4.7   Two approaches to interstellar solar sailing   50

Chapter 5
5.1   How spaceprobes shrink as a function of time.   59
5.2   An approach to the manufacture of a nanocomponent   60

Chapter 6
6.1   Schematic operation of a generalised nuclear rocket   67
6.2   Schematic operation of a nuclear-electric rocket   69
6.3   Two approaches to nuclear-pulse propulsion   72
6.4   An inertial electrostatic confinement fusion reactor   75

xviii List of illustrations

6.5	The beam-core engine.	78
6.6	A penning trap.	79

Chapter 7

7.1	Aspects of beamed-energy sailing.	84
7.2	The magsail as an interstellar braking mechanism.	87
7.3	A typical magsail deceleration profile.	89
7.4	The application of thrustless turning.	90
7.5	Thrustless electrodynamic turning.	91
7.6	A perforated solar sail.	93
7.7	A Fresnel lens in the outer Solar System.	95
7.8	Accelerating an interstellar spacecraft by momentum transfer.	97

Chapter 8

8.1	The proton-fusing interstellar ramjet.	102
8.2	Proton-fusing interstellar ramjet acceleration versus velocity.	104
8.3	A ram-augmented interstellar rocket.	105
8.4	A laser ramjet.	109
8.5	The ramjet runway.	110
8.6	The toroidal–magnetic ramscoop.	113

Chapter 9

9.1	Current produced in a conductor in motion relative to a magnetic field.	119
9.2	A double pendulum designed to demonstrate currents.	120
9.3	Two methods of generating unidirectional current flow in space.	120
9.4	Two hypothetical ZPE-propelled starships.	123

Chapter 10

10.1	An optical interferometer.	136
10.2	The use of an occulting disk.	136
10.3	The Hertzsprung–Russell diagram.	141

Chapter 11

11.1	The O'Neill Model III Space Habitat.	152

Tables

2.1	Solar System planet Bond albedos.	17
3.1	Properties of a carbon microtruss sail coated with an aluminium reflective layer.	31
4.1	Specific impulses of some chemical propellants.	38
4.2	Periapsis parabolic velocities and angular deflections for an unpowered Jupiter flyby.	43
6.1	Values of mass–energy conversion efficiency.	68

7.1	Fractional spectral reflectance, absorption and transmission of an aluminium light sail	94
8.1	Estimated RAIR mass ratios compared with Powell (1975) results	108
8.2	Performance of a fusion-ramjet runway compared with a fusion rocket	111
10.1	Confirmed 'normal extrasolar solar systems'	139
10.2	Confirmed 'close Jupiters'	139
10.3	Confirmed 'eccentric Jupiters'	140
10.4	The nearest nearby stars that might possess habitable planets	142

Introduction

The whole world? The whole Universe? Overhead, obscurity unveiled a star. One tremulous arrow of light, projected how many thousands of years ago, now stung my nerves with vision, and my heart with fear. For in such a Universe as this what significance could there be in our fortuitous, our frail, our evanescent community?

Olaf Stapledon, *Star Maker* (1937, 1968)

To H. G. Wells, Jules Verne, Olaf Stapledon and the other early masters of science fiction, the Universe was a place of wonder, majesty and infinite possibilities. Human desires and motivations seemed dwarfed to insignificance by the sheer scale of Creation. Today, with a few decades of space exploration under our collective belts, many people take a somewhat different approach. Yes, the cosmos is huge and our Earth is small, and an individual human lifespan is as nothing when compared with the billion-year evolutionary time frames needed to understand the life of a star or a galaxy.

But we have walked on the Moon, and our probes have landed on two neighbouring worlds and orbited or flown by others. In fact, four tiny emissaries of humanity have left the realm of the planets and will drift between the stars for near-eternities of time. Humans are children of the Universe, with as much right to exist as anything or anyone else. As described by John Lewis in 1996, there are untold riches in space, and we can find them. Minerals and isotopes from the Moon, asteroids or comets might revolutionise our lives in centuries to come. Strange lifeforms beneath the dry soil of Mars and the frozen oceans of Jupiter's moon Europa may forever alter the way we relate to the Universe or other species. One day we will almost certainly encounter our equals or superiors in the vast realm of Galactic space.

But before we can grab for these riches in marvellous spacecraft that future generations may construct, we must conjecture and design what Nigel Calder referred to as 'spaceships of the mind'. That is the function of this book – to acquaint would-be designers of deep-space ships with the techniques of their craft and accomplishments of their predecessors. But before we can consider our first 'mental spaceships', we must have a reasonable picture of the scale of that immense new theatre for human activity: the cosmos.

The scale of space

It is very easy to become daunted by the distances between the planets, let alone the enormously greater distances between the stars. Perhaps a good first step is to consider distances to our nearest celestial neighbours and relate them to common human concepts of space and time.

Consider first our lovely and clement planet Earth, which is a convenient starting point for our cosmic voyage. The equatorial radius of our nearly spherical planet is about 6,400 km. We have sailed around our planet's 40,000-km circumference in about three years, and circled it by balloon in less than a month.

The fastest operational commercial airliner – the Anglo-French Concord(e) – can cruise at about 2,000 km per hour. Since such supersonic airliners must stop for refuelling every few hours, a Concorde would require rather more than one day to circumnavigate the world.

But we can of course do better than this. If we leave Cape Canaveral onboard a Space Shuttle, we will be injected into low Earth orbit (LEO) with a velocity of about $8 \, km \, s^{-1}$. Spacecraft in LEO require about 90 minutes to complete one orbit and circumnavigate the globe.

Our nearest celestial neighbour – our planet's one natural satellite, the Moon – has a radius of about 1,700 km and a mass 1/81 that of the Earth. The Moon requires about 29 days to orbit the Earth in an elliptical path at an average distance of about 380,000 km, or 30 Earth diameters.

If we add another propulsive stage to our LEO spacecraft, we can leave Earth orbit at about $11 \, km \, s^{-1}$ on the track pioneered by the Apollo astronauts. If directed towards the Moon, such an interplanetary craft slows as it climbs higher from LEO. The ship's kinetic energy is traded for potential energy. Current-technology spacecraft require a few days to reach the Moon, and about a week for a round trip.

The nearest planet to Earth is Venus, which is sunward of our planet. At their closest to each other, Venus and Earth are separated by about 42 million km, but because of Venus' enormous atmospheric pressure and surface temperature, humans will not follow our robots to the surface of this forbidding world in the near future. But if our interplanetary spacecraft passes the Moon and follows a long, looping 'minimum-energy ellipse' to Venus, it will arrive in that planet's vicinity after about six months.

Mars is more interesting, and is a more suitable target for an early expedition by humans. With a thin atmosphere, frozen water in its polar caps and possible fossil life, this planet approaches the Earth to within about 75 million km. Our interplanetary spacecraft can depart LEO and reach the environs of Mars in about nine months. Leaving time for the planets to realign and some martian surface rest and recuperation for the crew, a round-trip expedition to Mars using current technology will have a duration of about three years. During the sixteenth century, Magellan's crew sailed around the Earth in about three years.

Since Mars is about the limit for human expeditions using current technology, our hypothetical Mars expedition might be an appropriate place to introduce a new distance unit: the Astronomical Unit (AU). An AU is about equal to 150 million km – the average separation between the Sun and the Earth. The average distance of Mars from the sun is 1.52 AU.

A photon of light emitted from our Sun's surface and travelling at 300,000 km s^{-1} takes about eight minutes to reach the Earth, and an additional four minutes to reach Mars. Our spacecraft on its nine-month one-way journey to Mars no longer seems so fast.

Exercise Intro. 1 Imagine a spacecraft capable of moving in a straight line between planets at 10 km s^{-1}. Estimate the minimum one-way travel time for the craft to reach Venus or Mars, starting from Earth, not including the time required for acceleration and deceleration.

Beyond Mars we first find the orbits of most asteroids – those rocky, mountain-sized remnants of early Solar System formation. Then we come to the realm of the giant planets: Jupiter, Saturn, Uranus and Neptune. Neptune's average distance from the Sun is about 30 AU. The last of the known Solar System worlds, tiny Pluto, orbits the Sun every 250 years in an eccentric orbit with an average distance from the Sun of 39.4 AU.

But planets and asteroids are not the entire story. Beyond the farthest planet and extending tens of thousands of Astronomical Units into space is the home of those celestial icebergs we call comets. In the vastness of extraplanetary space, even the 150-million km AU becomes infinitesimal.

The fastest craft launched by humans is Voyager 1, which used a gravity-assist manoeuvre at Jupiter to leave the Solar System at about 3.5 AU per year (about 17 km s^{-1}). The nearest stellar neighbour to our Sun, the triple star α/Proxima Centauri, is about 270,000 AU from the Sun. If Voyager were vectored on that star (which it is not) that intrepid robot would pass through the α Centauri system in approximately 77,000 years. But light traverses about 60,000 AU per year. A photon emitted by our Sun reaches α Centauri in about 4.3 years, and the star system is therefore 4.3 light years from the Sun.

Even the light year pales as we move far into interstellar space. Our Milky Way Galaxy contains a few hundred billion stars, and is about 100,000 light years in diameter. Neighbouring galaxies are separated by hundreds of thousands or millions of light years. The Universe, which contains billions of galaxies, is perhaps 20 billion light years across.

Exercise Intro. 2 Calculate the diameter of the Milky Way Galaxy in Astronomical Units. Calculate the breadth of the entire Universe in Astronomical Units.

These huge universal distances need not concern us now, as for a very long time to come, a radius of 10 or 12 light years from the Sun will be the limit of human exploration. If we can learn to cross these still-enormous distances in timescales of decades or centuries instead of millennia, many solar systems will open for us, even in this tiny fragment of universal immensity.

BIBLIOGRAPHY

Calder, N., *Spaceships of the Mind*, Viking, NY (1978).
Lewis, J. S., *Mining the Sky*, Addison–Wesley, Reading, MA (1996).

1

Motivations for deep-space travel

If man goes down I do not believe that he will ever again have the resources or the strength to defend the sunflower forest and simultaneously to follow the beckoning road across the star fields. It is now or never for both, and the price is very high.

Loren Eisley, *The Invisible Pyramid* (1970)

When trying to predict the future of spaceflight, we are faced with the same dilemma experienced by visionaries in all fields. In order to model what humans may (or may not) do at some point in the indeterminate future, we must examine the nature of what makes humans move along certain paths. Always uncertain, such 'futurology' is even more difficult today as human civilisation seems poised on the threshold of a transformation from independent nation states to a global entity. What might encourage this emerging global entity to expand its realm of operations from the friendly, solar-heated spaces near the Earth to the alien, frigid void beyond the planets and between the stars?

Many authors have considered this question in works of fiction and fact. Some suggest that the answer is to be found in history. Almost 3,000 years ago, the Ionian descendents of the Minoan/Mycenean Bronze Age civilisation, centred in the Aegean, began to expand beyond the eastern shores of that sea into the wild steppes of modern-day Russia. Before 600 BC, one Ionian city alone (Miletus) had established more than 60 colonies along the shores of the Black Sea. During the same era, other Dorian Greek cities were colonising Sicily and Southern Italy, founding cities that became the nuclei of modern-day Syracuse and Naples.

At about the same time, the Polynesians were beginning their epic saga of colonisation among the wide-flung Pacific islands. These people crossed a much greater expanse of ocean than did their Greek contempories and, incredibly, they accomplished this exploration without the advantage of literacy. Instead of written sailing instructions, Polynesian navigators crossed the watery void with the aid of epic poems that informed them about star positions and wind direction.

Can these terrestrial sociological models and others serve us in predicting the future of space travel? Probably not, unfortunately. No matter how much modern humans may wish to escape their overcrowded world in the manner of our ancient forebears migrating from overcrowded islands and city states, we are frustrated by the fact that the nearest Earth-like worlds are many trillion kilometres distant. Non-robotic space explorers and colonisers must therefore take a little of Earth along, simply in order to survive. The duration of voyages to such oases in space will be measured in centuries, and the cost of even a modest venture to another star will approximate the current US Gross National Product. Interstellar (or even interplanetary) colonisation seems destined never to play a role in resettling Earth's destitute billions or relieving the world population explosion. Only in that most distant future when an aging, swelling Sun promises to swallow Earth and the rest of the inner Solar System will there be sufficient incentive for interstellar colonisation by a significant fraction of the terrestrial population.

If interstellar colonisation cannot serve as a near-term driver to propel us into the void, what about exploration? We all know about the Golden Age of Exploration and similar eras, when curiosity propelled humans far from their native lands. And we all realise that the possibility of life on Mars (albeit extinct and primitive) might be the spark that ignites an era of human missions to the Red Planet during the twenty-first century. Perhaps the desire to explore the biological possibilities of distant comets might be sufficient to propel human-occupied ships to the outer fringes of the Solar System and beyond.

Sadly, however, advancing technology reveals that interstellar exploration by humans may also not soon emerge from the science fiction epic. Why send a human crew on a voyage that might consume decades or centuries if we can gain the same information from a much less massive robot that is more reliable and tolerant to loneliness than is a human, and not as prone to boredom? The very success of robots such as Voyager 1/2 and Mars Pathfinder/Sojourner mitigates against sending human crews on voyages of exploration far into the void.

If the urges to colonise or explore new lands are not sufficient to extend the range of human activities far beyond the Earth, what then of commerce? Might we mine the asteroid belt, the moons of Jupiter and the frozen comets of the Oort Cloud for materials precious to human civilization? Might trade ships criss-cross the Solar System in the twenty-second century in the same manner that sailing ships circled the globe in centuries gone by? And might these same ships be the nucleus of a society adventurous enough to venture beyond the Sun?

Many authors – beginning in the 1970s with the work of the late Gerard K. O'Neill – have now considered the potential role of space resources in the development of a space-faring civilization based in near-Earth or cislunar space. Although we have just begun to inventory the resources of our Solar System, it already seems apparent that space traders need not venture that far from Earth. Why mine a main-belt asteroid or an even more distant comet when all the resources we will need for millenia are to be found beneath the lunar surface or among the near-Earth objects (NEOs), those asteroidal and cometary bodies that approach the Earth closely?

Therefore, it seems that desires for living space, pure knowledge and commercial gain may not be sufficient motives to justify near-term human exploration far from the Sun. But there still remains one very powerful motive that may ultimately trigger the expansion of global human civilization into the extraplanetary realm: survival.

For many decades, geologist and paleontologists puzzled over the phenomenon of mass extinction. According to the fossil record, terrestrial biological evolution is generally a slow and stable process. Lifeforms of disparate classes – trilobites, dinosaurs and mammals – develop slowly, mutate and radiate gradually to fill available ecological niches. But then Nature suddenly and unexpectedly deals a wild card! Something happens, and happens quickly – a planet-wide event that upsets the ecological equilibrium, rendering thousands of species extinct and offering the opportunity to survivors to 'take over the Earth'.

Clues to the nature of these recurrent planetary catastrophes were uncovered in the 1970s by a father–son research team from the University of California at Berkeley. Physicist Luis Alvarez and his geologist son Walter investigated a clay layer at Gubbio in Umbria, Italy, that separates the end of the Cretaceous era 65 million years ago and the beginning of the Tertiary. Much to their amazement, the iridium content in the sediments of the Cretaceous/Tertiary boundary was greatly elevated. Follow-up studies by the Alvarez team and others have indicated that this iridium enhancement is a worldwide phenomenon.

Iridium is not a common constituient of Earth's crust, but elevated levels of this element are found in meteorites, the remnants of cosmic asteroidal or cometary objects that occasionally impact the Earth. Many geologists were won over by the theory of a cosmic event that ended the reign of the dinosaurs at the termination of the Cretaceous, but not all were convinced. The evidence was at best circumstantial until a 'smoking gun' was discovered. This was the heavily eroded remains of a large impact crater in Yucatan, Mexico, that is approximately 65 million years old and was apparently produced by an object about 10 kilometres in diameter.

Computer models (and Hollywood special effects) have since been used to predict and simulate the effects of a large asteroidal or cometary impact upon Earth's ecology. Initial effects include a huge fireball followed by a hydrogen-bomb-like mushroom cloud (without the bomb's radioactivity). Firestorms would devour all vegetation within hundreds of kilometres of ground zero if the impact occurred on land. A more probable ocean impact would result in 500-metre high tsunamis that might cross the planet's oceans at speeds of 1,000 kilometres per hour, causing vast destruction and coastline modification.

Some have speculated that the force of a large impact might trigger volcanic eruptions and earthquakes along the fault lines where the Earth's crustal plates intersect, vastly increasing the devastation. But Earth's ecology might recover relatively quickly if it were not for the 'comet winter' effect. Huge quantities of dust raised by the firestorms would be carried by stratospheric circulation around the planet. This aerosol shroud might have an upper atmospheric residence time measured in years.

For a few years after a major cosmic impact, summer will be absent over much or most of the Earth due to the dust-weakened sunlight intensity. Vegetation will wither and die, followed by the mass death by starvation of the herbivores that feed on the plants. Carnivores will not survive for long as their food supply dwindles.

Most of the Earth's large animals disappeared with the end of the Creataceous, but the early mammals survived, perhaps because they were very small or perhaps because they could hibernate until the climate warmed. Much smaller than their dinosaur cousins and able to fly to isolated regions fortuitously protected from the worst of the devastation, some early birds also survived.

Fortunately for us survivors, terrestrial impacts by objects in the multi-kilometre size range are infrequent events, occurring at intervals of tens of millions of years. But impacts by objects up to a few hundred metres in diameter are far more frequent. We might expect such a 'city-killing' asteroid or comet nucleus to strike the Earth once per century, with an explosive impact roughly equivalent to that of a 20-megaton hydrogen bomb.

The most recent such event occurred near Tunguska, a sparsely-inhabited region of Siberia, in 1908. Human and property damage was minimal, but millions would have died if the 'aim' had been very slightly different and the cosmic projectile had struck Moscow.

The Earth has experienced a number of near misses since the Tunguska event, and we are probably due for another impact before the twenty-first century is very far advanced. Because of the growth in human population since 1908 and the consequent spread of habitation, we cannot expect to be as lucky next time.

But there is one major difference between the modern world and that of 1908 or our Cretaceous-era predecessors. For the very first time, a terrestrial species has evolved that is capable of mapping the smaller bodies of the Solar System, exploring them and diverting those that threaten the Earth. The desire to protect the Earth from nearby objects that threaten all humanity may prompt us to become an interplanetary civilization. A true interstellar capability may develop as our global civilization examines those small bodies on the fringes of the Solar System that feed the population of near-Earth objects.

1.1 AN INVENTORY OF NEAR-EARTH OBJECTS

Only a tiny fraction of NEOs have been detected, let alone studied in depth. John Remo has recently estimated that there are about 20 NEOs in excess of 5 km in diameter, roughly 400 in the 2 km range, 2,000 in the 1 km range, and about 6,000 with a diameter of around 0.5 km. He estimates that there are about 100,000 NEOs in the 0.1 km range.

The impact on the Earth of a 5-km sized NEO would create a global mass extinction such as the one that doomed the dinosaurs. Much more frequent Tunguska-sized events, equivalent in destructive power to a large hydrogen bomb, result from the impact of NEOs in the 0.1 km size range.

In his 1998 paper in *Journal of the British Interplanetary Society*, Gregory Matloff evaluates Planetary Society data from *NEO News* to investigate the source of the known NEO population fraction for which reasonably accurate orbits have been computed. As shown in Figure 1.1, more than half of the known NEOs have eccentricities in excess of 0.55, and about 30% have inclinations in excess of 30°.

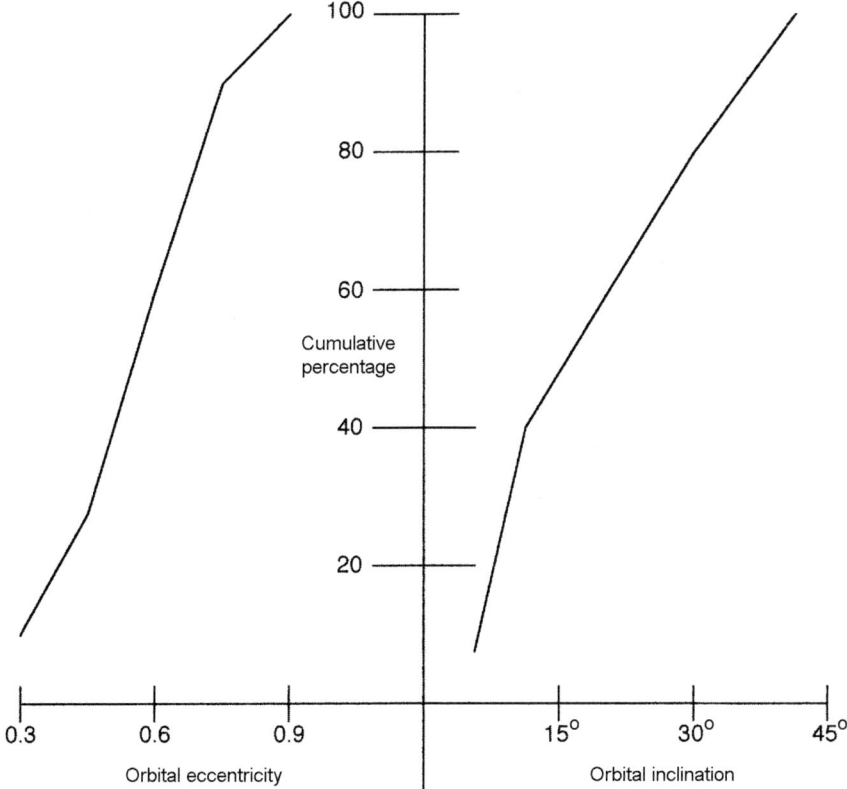

Fig. 1.1. Cumulative percentage of known near-Earth objects with orbital eccentricities and inclinations less than the indicated values.

Roughly 50% of known NEOs are in more elliptical orbits than most main belt asteroids; about 30% are in orbits more inclined to the ecliptic than are most main belt asteroids. We may reasonably conclude, therefore, that the NEO population is of both cometary and asteroidal origin.

If the Earth is threatened by the impending impact of a NEO of asteroidal origin, the best defence might be the launch of specially modified ballistic missiles. After rendezvousing with the offending object (hopefully in deep space beyond the orbit of the Moon), the missiles would be directed to ignite their nuclear warheads. Hopefully, the multiple nuclear detonations would be sufficient to nudge the NEO slightly from its previous path, thereby converting a bull's eye impact on the Earth into a near miss.

However, if the approaching NEO is a dark extinct comet nucleus, nuclear detonations might not be effective. Instead, in the manner of Comet Shoemaker–Levy 9 that impacted Jupiter in 1994, the cometary NEO might be 'calved' by the nuclear detonations into many radioactive fragments, each on an Earth-intercept trajectory. Much gentler solar-sail techniques might be necessary to gradually alter the trajectory of the offending NEO over the course of several decades.

To protect the Earth from approaching NEOs, we must very accurately determine the orbits of these objects and perhaps visit some of them to determine properties including composition and tensile strength. But since members of the NEO population have a dynamical lifetime much smaller than the lifetime of the Solar System, according to a paper by I.P. Williams, a terrestrial civilization truly intent upon protecting the Earth must investigate the origins of the near-Earth objects and perhaps deflect celestial bodies before they join this population.

1.2 CONSIDERING NEO ORIGINS

We may expect that most asteroids will long remain in their stable region between the orbits of Mars and Jupiter. But as J. Wisdom demonstrated in 1982, some asteroids approaching Jupiter might have their orbital eccentricities altered by the giant planet. Close encounters between an asteroid and Mars might further alter the objects trajectory, causing it to become a NEO. More recent work by Edward Belbruno and Brian Marsden has demonstrated that comets passing near Jupiter can also be perturbed to join the NEO population. (We expect that close encounters between comets and other giant planets might also feed objects into the NEO population). As discussed by Carl Sagan and Ann Druyan, perturbations to the Oort Cloud (a region of thousands of millions of comets extending outwards of a light year or more from the Sun) by a passing star could conceiveably deflect thousands of comets into the inner Solar System.

But the Solar System is a vast place. One might think that this very vastness might protect the Earth from a cometary or asteroidal object that has been deflected towards the Earth.

To investigate this supposition, we follow the logic of Gregory Matloff and Kelly Parks. Consider an object moving towards the Sun in a parabolic orbit with a velocity V_{para}. The mass of the Sun is M_{Sun}, and the coordinate system is directed positive outward from the Sun. Applying elementary kinematics, the object's velocity towards the Sun can be estimated from

$$V_{para} \approx -1.4(GM_{sun})^{1/2} R_{cent}^{-1/2} \qquad (1.1)$$

where G is the gravitational constant and R_{cent} is the object's distance from the Sun's centre.

If the orbital perturbation occurs at time $t = 0$ when the object is R_{init} from the Sun's centre and the object is R_{fin} from the Sun's centre at time t, the substitution $V_{para} = dR_{cent}/dt$ can be used to integrate equation (1.1):

$$\int_{R_{init}}^{R_{fin}} R_{cent}^{1/2} dR_{cent} \approx -1.4(GM_{sun})^{1/2} \int_0^t dt \qquad (1.2)$$

If it is assumed that $R_{init} \gg R_{fin}$, MKS values for G, and M_{sun} are inserted, and we substitute for the number of metres in an Astronomical Units (AU) and the number of seconds in a year, we obtain the time in years required for the object to fall to the inner Solar System from an initial position of $R_{\text{init},au}$ Astronomical Units from the Sun:

$$T_{\text{year}} \approx 0.077 R_{\text{init},au}^{3/2} \tag{1.3}$$

A comet or asteroid deflected into a parabolic orbit by Jupiter at 5.2 AU from the Sun requires less than a year to reach the inner Solar System. The warning time for planetary catastrophe might be small indeed!

Exercise 1.1 Validate all steps in the derivation of equation (1.3). Then estimate the times for comets deflected by Saturn, Uranus, or Neptune into parabolic solar orbits to reach the inner Solar System.

1.3 THE DIFFICULTY OF TELESCOPIC EXPLORATION OF NEOS NEAR THEIR POINT OF ORIGIN

As an alternative to robotic or human-occupied exploratory missions to NEOs near their point of origin, we might consider remote exploration using terrestrial or Earth-orbiting telescopes. This is very difficult, as revealed by the following calculation.

Since the amount of solar energy reaching the Earth per second per unit area (the solar constant) is about 1,400 W m^{-2}, the solar energy per second per square metre reaching a celestial object R_{au} from the Sun is about $1,400 R_{au}^{-2}$. If the object is spherical and has a fractional reflectivity of *REF* and radius of *RAD* metres, the amount of solar energy reflected from the object is $(REF)(\pi RAD^2)\, 1,400 R_{au}^{-2}$. If the object is a spherically-symmetric isotropic reflector and it is $R_{e,au}$ from the Earth, the reflected light energy per second from the object entering a near-Earth telescope of radius R_{tele} metres is calculated as

$$W_{\text{tele}} = 350\pi(REF)\left[\frac{(RAD)(R_{\text{tele}})}{1.5 \times 10^{11} R_{au} R_{e,au}}\right] \text{W m}^{-2} \tag{1.4}$$

The factor 1.5×10^{11} in equation (1.4) results from the conversion of Astronomical Units to metres. The telescope/celestial-object geometry assumed in the derivation of equation (1.4) is summarised in Figure 1.2.

We estimate the light energy per second entering a 10-m aperture telescope situated on or near the Earth from a 10% reflective celestial body situated 5 AU from both the Sun and telescope from equation (1.4) as 7.8×10^{-16} W. If the 10-m telescope is located outside the Earth's atmosphere – in which case extinction effects can be ignored – the reflected solar energy from the celestial object reaching each square metre of the telescope per second (also called the received flux, f_{rec}) is 2.49×10^{-18} W m^{-2}.

8 Motivations for deep-space travel [Ch. 1]

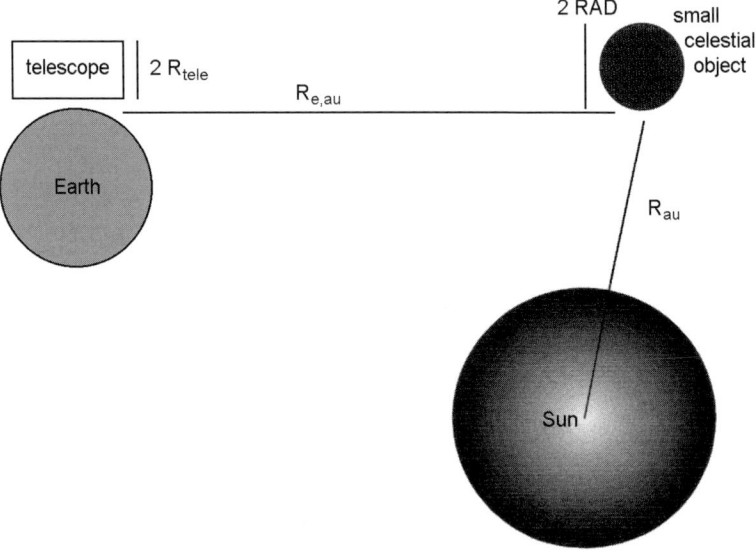

Fig. 1.2. Geometry for remote viewing of a small celestial object of reflectivity REF.

Exercise 1.2 Validate the derivation of equation (1.4). Calculate W_{tele} and F_{rec} for a number of telescope sizes, distance values, celestial object sizes and reflectivities.

From equation (3.1.13) of Kitchen's handbook, it is possible to relate the sensor-independent (or bolometric) apparent magnitude m_{bol} of any celestial object observed above the Earth's atmosphere to the received flux:

$$f_{rec} = 2.5 \times 10^{-8} \times 10^{-0.4 m_{bol}} \text{ W m}^{-2} \tag{1.5}$$

Substitution in equation (1.5) reveals that the apparent bolometric extra-atmospheric magnitude of our hypothetical object from which a flux of 2.49×10^{-18} W m^2 is received is about 25.

Although challenging, detection of 25th magnitude objects is routinely accomplished by instruments such as the Hubble Space Telescope. But the problem of telescopically resolving fine details on the surface of a small, distant object is another matter.

The angle θ subtended by our hypothetical small celestial object is equal to $(2RAD/R_{e,au}/1.5 \times 10^{11})$ radians. Applying the standard form of Rayleigh's criterion (as expressed in Chapter 1 of Kitchen's handbook).

$$\theta = \frac{2RAD}{R_{e,au} \times 1.5 \times 10^{11}} = \frac{1.22\lambda}{2R_{tele}} \tag{1.6}$$

where λ is the wavelength of the light received by the telescope from the celestial object. The object in question subtends about 2.5×10^{-9} radians, or approximately 0.0005 arcsec. Assuming 0.5 µm wavelength yellow light, a telescope aperture radius of approximately 100 m is required to resolve the celestial object in question. A space-based interferometer seems to be required, as opposed to a single-mirror telescope.

1.4 ROBOTIC EXPLORATION OPTIONS

As an alternative to telescopic observation from the vicinity of the Earth, our future space-faring civilization may desire to explore distant objects of cometary or asteroidal origin using robotic probes. Such probes must be very intelligent and highly autonomous because of the communication time delay caused by the speed-of-light limitation. At a separation of 5 AU, the trip time for a radio signal to the spacecraft is about 40 minutes.

One alternative to super-intelligent, autonomous miniaturised probes might be the location of a command centre onboard a peopled spacecraft located within a few hundred thousand kilometres of the object being explored. Robotic probes deployed by the larger craft could be under direct control as they explore the surface of the small celestial body under investigation.

1.5 DIRECTED PANSPERMIA

This possible motivation for robotic deep-space missions is technically feasible, although ethically disturbing. Our technology has evolved to the point at which we could launch robotic interstellar spacecraft on very long-duration missions. Requiring tens of thousand or hundreds of thousands of years to reach a neighbouring young star or more distant star-forming region, such craft could carry nano-payloads of freeze-dried genetically-adapted microorganisms. For a surprisingly low cost, terrestrial life could spread its DNA throughout the Universe.

The technology for such directed panspermia missions is nearly at hand. Acceleration of the tiny (micro-gram mass) panspermia capsules could be accomplished by current-technology solar-photon sails no more than a metre in diameter. Deceleration near a young star would utilise the solar-photon sail as a deceleration mechanism; deceleration in star-forming nebulae would be accomplished by matter drag.

As well as describing the physical and biological technology of panspermia missions, Michael Mautner has touched on the ethical considerations. A technological civilization threatened with extinction might elect to seed its nearest celestial neighbours so that the life of its planet can outlive the launching civilization and possibly also its planet. But individuals may reasonably oppose panspermia because it randomly spreads life through the Universe, and the revived microorganisms might destroy or severely modify young, naturally evolving ecosystems in the environments encountered.

Because of the rapid advance in terrestrial biotechnology and the near-term feasibility of solar sails capable of launching payloads on extrasolar trajectories, panspermia ethics is not a subject for the distant future. Unless international agencies debate the advisibility and desirability of spreading terrestrial DNA throughout the Universe, even such groups as foundations or small universities may unilaterally begin the process of directed panspermia in the early decades of the twenty-first century.

1.6 BIBLIOGRAPHY

Belbruno, E. and Marsden, B. G., 'Resonance Hopping in Comets', *Astronomical Journal,* **113**, 1433–1444 (1997).

Finney, B. R., 'Voyagers into Ocean Space', in *Interstellar Migration and the Human Experience*, ed. B. R. Finney and E. M. Jones, University of Chicago Press, Berkeley, CA (1985), pp. 164–179.

Hansson, A., *Mars and the Development of Life*, 2nd edn., Wiley–Praxis, Chichester, UK (1997).

Kitchin, C. R., *Astrophysical Techniques*, 2nd edn., Adam Hilger, New York (1991).

Lee, R. B., 'Models of Human Colonization: San, Greeks, and Viking', in *Interstellar Migration and the Human Experience*, ed. B. R. Finney and E. M. Jones, University of Chicago Press, Berkeley, CA (1985), pp. 180–195.

Matloff, G. L., 'Applying International Space Station (ISS) and Solar-Sail Technology to the Exploration and Diversion of Small, Dark Near-Earth Objects (NEOs)', in *Missions to the Outer Solar System and Beyond, 2nd IAA Symposium on Realistic Near-Term Scientific Space Missions*, ed. G. Genta, Levrotto & Bella, Turin, Italy (1998), pp. 79–86. Also published in *Acta Astronautica*, **44**, 151–158 (1999).

Matloff, G. L., 'The Near-Earth Asteroids: Our Next Interplanetary Destinations', *Journal of the British Interplanetary Society*, **51**, 267–274 (1998).

Matloff, G. L. and Parks, K., 'Interstellar Gravity Assist Propulsion: A Correction and New Application', *Journal of the British Interplanetary Society*, **41**, 519–526 (1988).

Mautner, M. N., 'Directed Panspermia. 3. Strategies and Motivation for Seeding Star-Forming Clouds', *Journal of the British Interplanetary Society*, **50**, 93–102 (1996).

Mautner, M. N. and Matloff, G. L., 'An Evaluation of Interstellar Methods for Seeding New Solar Systems', in *Missions to the Outer Solar System and Beyond, 1st IAA Symposium on Realistic Near-Term Scientific Space Missions*, ed. G. Genta, Levrotto & Bella, Turin, Italy (1996), pp. 273–278.

Mileikowsky, C., 'How and When Could we be Ready to Send a 1,000-kg Research Probe with a Coasting Speed of 0.3 c to a Star', *Journal of the British Interplanetary Society*, **49**, 335–344 (1996).

Ostro, S. and Sagan, C., 'Cosmic Collisions and Galactic Civilizations', *Astronomy and Geophysics*, **39**, 422–424 (1998).

Remo, J. L., 'An Approach to Assessing the Technological Cost/Benefit of Critical and Sub-Critical Cosmic Impact Prevention', *Journal of the British Interplanetary Society*, **51**, 461–470 (1998).

Sagan, C. and Druyan, A., *Comet*, Random House, New York (1985).
Williams, I. P., 'Sub-Critical Events: the Origin of the Impactors', *Journal of the British Interplanetary Society*, **51**, 445–450 (1998).
Wisdom, J. 'The Origin of the Kirkwood Gaps: A Mapping for Asteroidal Motion Near the 3/1 Commensurability', *Astronomical Journal*, **85**, 1122–1133 (1982).

2

The realms of space

Before brave explorers sail the seas in search of new and exotic lands, they ought first to prepare themselves by surveying their own country. Its properties may reappear elsewhere and help them make sense of new discoveries. One's native land might have hills and valleys, forests and deserts, rivers and lakes – features that a foreign land might also possess. Whether at home or abroad, these features would probably obey the same basic laws.

Ken Croswell, *Planet Quest* (1955)

Before humans commit either their robot emissaries or themselves to voyages deep into the abyss, it is worthwhile for them to review what is known about their home Solar System. Even though each set of stellar worlds we encounter will be unique and individual in some respects, the same laws apply throughout the cosmos. What we know of the Sun's retinue of worlds, satellites, asteroids, comets, fields and particles can serve as a model for what we encounter orbiting other stars.

2.1 THE QUESTION OF ORIGINS

Current thinking regarding the origin of our Solar System is outlined in many astronomy texts (see for example, Chaisson and McMillan (1996)). All evidence supports the hypothesis that solar-system formation is a natural consequence of star formation. Even before the recent observational discoveries of extrasolar planetary systems, astronomers were convinced that planetary systems are far from uncommon.

About 5,000 million years ago, a cool, rotating gas and dust cloud or nebula filled our region of the Milky Way Galaxy. Mostly hydrogen and helium with a small proportion of more massive elements and carbon dust particles, this nebula must have had dimensions measured in fractions of a light year and hydrogen-atom densities in the neighbourhood of 1,000 per cubic centimetre.

This primeval nebula may have existed for ages, and it may still have been present today if a nearby stellar catastrophe had not triggered star formation in its interior. Perhaps within it, or perhaps within a few light years of the primeval nebula, a massive star formed. Such stars evolve rapidly, moving away from the stable, hydrogen-burning main sequence after a period of millions of years. As it aged, the massive star expanded to become a supergiant that would engulf the orbit of Mars if it replaced the Sun. Having exhausted its hydrogen fuel, the massive star consumed helium, carbon and heavier elements as it converted more and more of its material to energy and heavy elements that sank, as an inert ash, to its core. For millions of years, the radiation pressue of the star's self-generated photons balanced gravity and prevented it from collapsing.

Finally, the fuel was gone and the huge star began to collapse. Temperatures and pressures rose in the inert core as the outer layers fell inward. A whole new series of thermonuclear reactions suddenly became possible in the core, and the star rebounded in an enormous explosion. In the twinkling of an eye, a mass equal to thousands of Earths was converted into energy as the nuclear furnace of the dying star consumed elements as massive as iron, to produce all elements in the Periodic Table up to and including uranium. For a brief time, before it faded to oblivion, the light emitted by the supernova might have challenged the combined luminous output of all stars in the Galaxy.

The nebula would have experienced this stellar death in two ways. First, streamers of heavy elements produced by the supernova would have 'doped' the nebula with elements as massive as uranium. Second, the expanding gases – previously part of the supergiant's outer layers – would turbulently mix with the substance of the nebula.

Exercise 2.1 Consider an uncollapsed spherical interstellar cloud with an initial density of 10^8 hydrogen atoms per cubic metre (Butler *et al.*, 1978). Show that the cloud's radius is $\approx 10^{13}$ km (about 1 light year) if it has the Sun's mass (2×10^{30} kg).

Turbulent eddies would have rotated throughout the primeval nebula, perhaps resembling slow-motion whirlpools. According to Heiles (1976), a portion of an interstellar cloud will collapse to a flattened disc resembling the primeval Solar System if centrifugal acceleration at its edge is larger than or about equal to gravitational acceleration:

$$\omega_{cloud}^2 R_{cloud} \geq \frac{GM_{cloud}}{R_{cloud}^2} \tag{2.1}$$

where ω_{cloud} is cloud-portion angular velocity, R_{cloud} is cloud-portion radius, M_{cloud} is cloud-portion mass and G is the gravitational constant. As the cloud-portion collapses to form the disc of an infant solar system, angular momentum (per mass) $\omega_{cloud} R_{cloud}^2$ is conserved. Initial cloud-portion angular momentum is transferred to rotation of the protostar and rotation/revolution of the protoplanets. Conservation of angular momentum is why most Solar System planets rotate around their axis in the same direction as they revolve around the Sun.

Exercise 2.2 To demonstrate the effect of angular-momentum transfer from protostars to protoplanets, consider that the cloud in Exercise 2.1 collapses to form a Sun-like star with a radius 7×10^5 km that does not transfer its angular momentum to its planetary system. Assume that the uncollapsed cloud's rotational velocity is 0.5 km s^{-1} (about equal to its internal velocity dispersion of 0.5–1 km s^{-1} (Larson, 1973)). First equate cloud and protostar angular momenta using $\omega_{cloud} R^2_{cloud} = \omega_{protostar} R^2_{protostar}$. Then calculate the protostar's rotational velocity (v_{rotate}) using $\omega_{protostar} = v_{rotate}/R_{protostar}$). Compare this protostar's rotational velocity with that of the Sun, which can be calculated by dividing the Sun's circumference by its rotational period (about 30 days). What a difference a planetary system makes in a star's rotational velocity. The star in question could not survive unless it transferred most of its angular momentum to something!

Transfer of an interstellar cloud's angular momentum to protoplanets is not the only change in the nature of the collapsing cloud. Temperature, pressure, and density build up in the central condensation – the protostar – until that object is able to sustain thermonuclear burning of hydrogen (with helium and energy as the reaction products) in the region near the infant star's core. Collapse ends as the radiation pressure of the released energy counterbalances the self-gravitation of the material collapsing towards the new star's centre.

Emitted radiation and a stellar wind flow from the infant star. These factors contribute to the evaporation of nebular dust and gas from the inner reaches of the new solar system. The primeval hydrogen/helium atmospheres of the inner planets (Mercury, Venus, Earth and Mars in the case of our Solar System) would also evaporate (over a period of thousands or millions of years) under the influence of the young star's emissions.

2.2 REALMS OF FIRE, WATER AND ICE

As the young star settles down to begin its long career as a stable, main sequence, hydrogen-burning star, fragments of ice and rock (planetesimals) cruise the inner Solar System, occasionally impacting the inner planets. Oceans form from these impacts, as do primeval, non-oxidising atmospheres.

However, when we survey the inner planets of our Solar System we find a wide variety of atmospheric environments. Tiny Mercury is a cratered desert, greatly resembling the Moon. Venus, once thought to be a near-twin of the Earth, bakes beneath a high-pressure, highly acidic carbon dioxide atmosphere. Dry, frozen Mars shows tantalising hints of past wet eras when surface life may have been present. Only our Earth, with its clement oxygen-rich atmosphere and abundant oceans, is a fitting abode for widespread life.

As we survey any other planetary system for signs of life, the patterns of our own Solar System can serve as a model. Although planet size, mass, geology and atmosphere play significant roles in determining habitability, the dominant factor is the radiant flux received by a planet from its primary star. As defined by Dole and Asimov (1964), an ideal region to search for life-bearing worlds around a distant star is the 'ecosphere' – the region bounded on the inside by the point at which an Earth-like planet's oceans would boil, and on the outside by the point at which the oceans would freeze. Widespread life on a planet's surface could exist only in the intermediate 'water' zone, bounded on the inside by a realm of 'fire' and on the outside by a realm of 'ice'. Dole has calculated that the inner ecosphere boundary for a G0 star like the Sun is about 0.9 AU, and the outer boundary is about 1.2 AU. A brighter F0 star has an ecosphere 1.8–2.6 AU from the star's centre, and the ecosphere of a dimmer K0 star is 0.60–0.66 AU from the primary star.

Many authors have expanded upon and refined Dole's early work on ecosphere limits. For example, Kasting *et al.* (1993) more conservatively calculated the Sun's continuously habitable ecosphere dimensions as 0.95–1.15 AU. Wiegert and Holman (1997) have argued, using computer simulations, that certain binary stars, including α Centauri (the sun's nearest interstellar neighbour) could have stable planetary orbits within the ecosphere of both stars.

The ecosphere's boundaries may not be constant over long periods of time. Hansson (1997) reviews arguments that Mars' surface life may have formed early in the Solar System's history. Direct robotic probing and eventual peopled expeditions during the next few decades should reveal whether fossilised life is present on that hostile world, or whether martian life has even retreated underground.

Ecospheric boundaries are subject to expansion if we consider non-solar sources of heat. As reviewed by Chapman (1999), observations from the Galileo probe have revealed that Europa – a satellite of Jupiter more than 5 AU from the Sun – may possess a liquid water ocean with greater volume than that of the Earth. A likely heat source for this possible abode of life is tidal interactions between Moon-sized Europa and giant Jupiter.

A small minority of bioastronomers have argued that life in the Universe is absolutely ubiquitous. According to Wickramsinghe, Hoyle and their colleagues, exobiologists should even consider interstellar clouds and comets as possible abodes of bacterial life. Davies *et al.* (1985) present the 'majority-view' counter-argument to this thesis.

2.3 SOLAR RADIANT FLUX AND PLANET EFFECTIVE TEMPERATURE

To estimate a planet's potential habitability, the first step is to determine the planet's effective (black-body) temperature, due to the radiant flux from the planet's star incident upon the planet and absorbed. In this calculation, we follow the reasoning of Brandt and Hodge (1964).

The solar flux incident upon a surface normal to the Sun – the solar constant – is defined as

$$S_c = \frac{1,400}{R_{au}^2} \text{ W m}^{-2} \tag{2.2}$$

Table 2.1. Solar System planet Bond albedos

Terrestrial planets	Albedo	Jovian planets	Albedo
Mercury	0.06	Jupiter	0.73
Venus	0.71	Saturn	0.76
Earth	0.33	Uranus	0.93
Mars	0.17	Neptune	0.84

Source: Goody and Walker (1972)

where R_{au} is the distance to the Sun, in Astronomical Units. The factor '1,400' in equation (2.2) will vary with the primary star's spectral class. According to Dole (1964), this factor should be replaced by 6,240, 2,900, 735, 395, 200, 83 if the G0-class Sun were respectively replaced by an F0, F5, G5, K0, K5 or M0 main-sequence star.

The amount of radiant energy per second absorbed by a near-spherical planet with radius R_{planet} metres is expressed as

$$P_{planet,absorbed} = (1 - BA) S_c \pi R_{planet}^2 \text{ W} \tag{2.3}$$

where BA is the Bond albedo of the planet (the amount of light reflected in all directions/amount of light incident on planet). The last term in equation (2.3), πR_{planet}^2, is the planet's cross-sectional area. The Bond albedos of the Earth and other major planets in our Solar System are listed in Goody and Walker (1972) and in many other sources, and are reproduced in Table 2.1.

Because the planet is a spherical object, the radiant flux emitted by the planet can be expressed as

$$W_{planet,emit} = (1 - BA) \frac{S_c}{4} \text{ W m}^{-2} \tag{2.4}$$

By next applying the Stefan–Boltzmann law, the planet's black-body or effective temperature can be expressed as

$$T_{eff} = \left[\frac{(1 - BA)}{4\sigma} S_c \right]^{1/4} \text{ Kelvin (K)} \tag{2.5}$$

where σ is the Stefan–Boltzmann constant (5.67×10^{-8} W m^{-2} K^{-4}).

If we next substitute into equation (2.5) the value of Bond albedo for the Earth from Table 2.1, the definition of the solar constant from equation (2.2) for the Earth at 1 AU from the Sun and the numerical value of σ, we estimate the value of Earth's effective temperature as 253 K.

Exercise 2.3 Calculate the effective temperatures for Venus (at 0.72 AU from the Sun) and Mars (at 1.52 AU from the Sun). Then repeat these calculations to determine what would happen to effective planet temperatures if the Sun were replaced with an F5, G5, or K0 star.

Note that the Earth's effective or black-body temperature is well below the freezing point of water, 273 K. The fact that the Earth has liquid oceans at all has a great deal to do with the planet's atmosphere, as discussed next.

2.4 THE EFFECT OF ATMOSPHERIC OPTICAL DEPTH

Planetary temperatures as calculated above are increased by the radiant absorption effects of planetary atmospheres. Radiant absorption in a planet's atmosphere is characterised by the optical thickness (OT). According to Kondratyev (1969), OT for any wavelength of light at a level z in a planet's atmosphere is the integral over optical path length between z and infinity of the product of atmospheric attenuation coefficient for that wavelength and the air density.

From Jastrow and Rasool (1965), a planet's surface temperature is related to its atmospheric infrared optical thickness and the effective black-body temperature by

$$T_{\text{surface}} = \left(1 + \frac{3}{4} OT_{ir}\right)^{1/4} T_{\text{eff}} \text{ K} \qquad (2.6)$$

Jastrow and Rasool list the infrared optical thickness as 0, 55.4, 1.4, 0.6 and 4.3 for Mercury, Venus, Earth, Mars and Jupiter. Substituting equation 1.4 for Earth's infrared optical thickness and our previously calculated effective black-body temperature of 253 K into equation (2.6), Earth's average surface temperature is calculated to be about 300 K.

Exercise 2.4 Calculate the surface temperature of Venus and Mars using the results of Exercise 2.3, the optical thicknesses presented above, and equation (2.6).

Physically, the temperature-increasing effects of a planet's atmosphere is identical to the Greenhouse Effect. Various infrared-absorbing gases in a planet's atmosphere (CO_2, H_2O, O_3, CH_4 and NO_2 for the Earth and CO_2 for Venus) trap reradiated infrared radiation from the planet's surface, thereby increasing the atmospheric temperature. The Greenhouse Effect on Venus is due to natural causes. A portion of global warming on the Earth is due to anthropomorphic effects, particularly CO_2 emissions by fossil fuels.

2.5 THE LIFETIME OF A PLANET'S ATMOSPHERE

Nothing is forever – not even the atmosphere of a large planet. We can crudely estimate the lifetime of a constituent of a planet's atmosphere following the arguments of Jastrow and Rasool, conservatively assuming that the interplanetary medium and asteroidal/cometary impacts do not increase that constituient's atmospheric concentration after the planet's formation.

We first define the scale height of the atmosphere in the exosphere (the boundary between the atmosphere and space):

$$H_{ex} = 10^{-3} \frac{kT_{ex}}{m_{con} g} \text{ km} \tag{2.7}$$

where k is the Boltzmann constant (1.381×10^{-23} Joule/degrees K), T_{ex} is the exospheric temperature in degrees K, m_{con} is the atmospheric constituent's molecular mass in kg, and g is the planet's gravitational acceleration in the exosphere, in m s^{-2}.

For the Earth, the exospheric temperature is about 1,500 K. Since the Earth's exosphere is only a few hundred kilometres above the ground, $g = 9.8$ m s^{-1}. The scale height of atomic oxygen (with an atomic mass of 16 and $m_{con} = 16 \times 1.67 \times 10^{-27} = 2.67 \times 10^{-26}$ kg) is therefore 79 km. For planets other than the Earth, $g = M_{planet} G / R_{planet}^2$, where M_{planet} is the planet's mass, R_{planet} is its radius, and G is the gravitational constant.

Exospheric temperatures are estimated by Jastrow and Rasool for planets other than the Earth. For Venus, Mars, and Jupiter, T_{ex} is respectively about 2,600 K, 1,100 K and 130 K.

Exercise 2.5 Estimate g at the surface of a planet with twice the Earth's mass and twice the Earth's radius. Then calculate exospheric scale heights for atomic oxygen and atomic hydrogen (atomic mass of 1) for this planet if $T_{ex} = 1,100$ K and 1,500 K.

From statistical mechanics, Jastrow and Rasool next define the mean thermal velocity of molecules of the atmospheric constituent in the planet's exosphere:

$$V_{con,av} = 2 \left(\frac{2kT_{ex}}{\pi m_{con}} \right)^{1/2} \text{ m s}^{-1} \tag{2.8}$$

The approximate lifetime of the gas constituent in the planet's exosphere is defined next by Jastrow and Rasool as

$$t_{ex} \approx \frac{4H_{ex}}{V_{con,av}} \frac{\exp(R_{ex}/H_{ex})}{R_{ex}/H_{ex}} \text{ s} \tag{2.9}$$

where R_{ex} is the planet's exospheric radius.

Exercise 2.6 Earth's exosphere will not escape into space in the near future! First use equation (2.8) to validate that the mean thermal velocity for exospheric atomic oxygen is about 36 m s^{-1}. Then substitute $H_{ex} = 79$ km and $R_{ex} \approx 6,500$ km into equation (2.9) to estimate the lifetime of atomic oxygen in Earth's exosphere as about 6×10^{25} year. Then repeat the exercise for atomic hydrogen.

Equation (2.9) actually underestimates a planetary atmosphere's lifetime, because most of a planet's atmosphere is below the exosphere. Unpublished corrections to equation (2.9) by Richard Stewart, which are cited by Jastrow and Rasool, reveal that the lifetime of a planetary atmosphere constituent is typically about 10^6 times greater than indicated by equation (2.9) for terrestrial planets Venus, Earth, and Mars.

2.6 COMPARITIVE PLANETOLOGY: AN APPRECIATION OF THE LIFEZONE

As outlined in Figure 2.1, an understanding of equation (2.9) allows us to appreciate why some worlds bear life and others do not. A large exospheric temperature results in a large scale height. Because the exponential factor in the equation is the most rapidly varying, high exospheric temperature will result in a more rapid escape of a planet's atmosphere. If the planet is massive or if we consider very massive atmospheric gases, scale height decreases, the exponential factor rapidly increases and the atmosphere's lifetime is increased. A gas giant world such as Jupiter or Saturn will have a large exospheric radius, which rapidly increases the exponential factor and the atmosphere's lifetime. Consider next a nearly airless world such as the Moon or Mercury. Since all the planet's minuscule atmosphere is exosphere, the correction factor that increases atmosphere lifetime in equation (2.9) should not be applied.

To increase the rigour of the preceeding analysis and include worlds such as Europa, some analytical modification to account for non-solar planetary-surface heating effects is required. This could be accomplished by including an additional

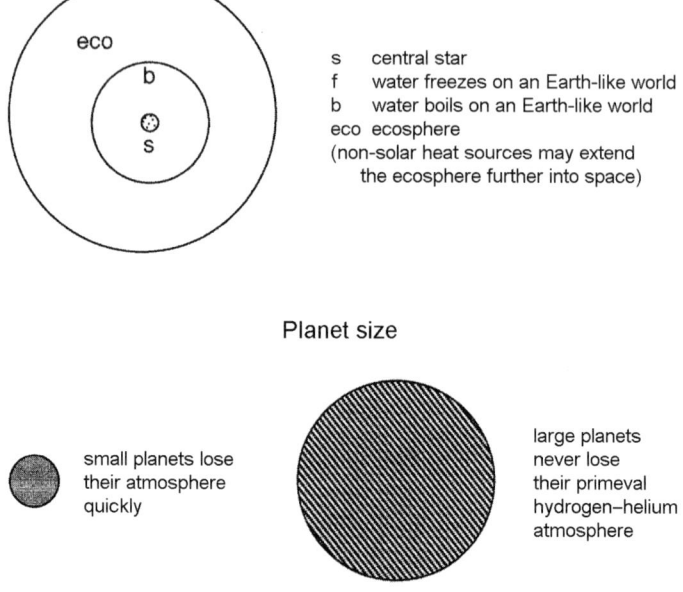

Fig. 2.1. Factors influencing a planet's habitability.

temperature increment term in equation (2.6) to account for tidal, tectonic or local geothermal heating of a planet's surface.

One final aside is presented to further emphasise the power of planetary-atmosphere lifetime calculations. In the early 1990s the author was asked to consult on a science fiction novel co-authored by Apollo 11 astronaut Buzz Aldrin (B. Aldrin and J. Barnes, *Encounter with Tiber*, (1996). For plot purposes, the authors required the existence of jovian planets orbiting about 1 AU from the two main α Centauri suns. Doubting the long-term lifetime of a hydrogen/helium atmosphere that close to a Sun-like star, the author applied the arguments leading to equation (2.9); and much to his surprise, learned that such a close-giants atmosphere is stable for billions of years. Although these results were never published, they effectively predicted the discovery later in the decade of 'hot Jupiters' orbiting Sun-like stars (discussed further in Chapter 10).

2.7 BEYOND THE PLANETS: THE REEFS OF SPACE

Although one probe from Earth (Mariner 10) has flown by Mercury and a few craft have visited the outer planets, most of our exploration to date has been within the vicinities of Venus, Earth and Mars. As we venture farther from the ecosphere of the Sun, the environments we encounter become more and more alien. Moving farther from the Sun, we first come to the giant planets Jupiter, Saturn, Uranus and Neptune. With their primeval atmospheres largely intact, these cold and distant giants might be considered laboratories of the Solar System's early history.

Located mostly between Mars and Jupiter, but with members closer to the Earth and farther from the Sun, is the class of sub-planetary objects called the asteroids. Asteroids are present in rocky, stony and carbon-rich varieties, and some have smaller satellites. Although their origin is far from universally agreed upon, it seems likely that they are remnants of a failed planet that could not coalesce because of the proximity to giant Jupiter. Like the major planets, most asteroids orbit the Sun close to the same plane – the ecliptic. The study of asteroids is becoming significant because of the realisation that these objects occasionally smash into the Earth, with dire consequences.

Between about 30 and 50 AU from the Sun is the Kuiper Belt, a region of frozen bodies consisting mostly of water, methane and ammonia ices, and with typical sizes of 100–1,000 km. Although our study of the Kuiper Belt has barely begun, astronomers expect thousands of bodies to inhabit this region. Emma Bakes (2000) of NASA Ames Research Center has elaborated the scientific justifications for the near-term *in situ* exploration of the Kuiper Belt. The largest known Kuiper Belt object, the planet Pluto, orbits the Sun in the most eccentric orbit of all the planets at a mean distance of 39.3 AU from the Sun. Pluto's orbit is inclined 17.2° to the ecliptic – more than any other Solar System planet. Accompanied by its giant satellite Charon, Pluto is a tempting target for early twenty-first century space-mission planners.

Beyond the Kuiper-Belt 'cometoids' is the realm of the true comets – the Oort Cloud. Although comets heat up and exhibit 10 million km tails when they pass close to the Sun, these Oort Cloud residents spend most of their time as frozen icebergs, with dimensions of about 10–20 km. With a central rocky nucleus and alternating layers of (methane, ammonia and water) ice and dust, the trillion or so Oort Cloud comets spend most of eternity in a spherical halo extending perhaps 100,000 AU from the Sun. Comets might be directed sunward by the influence of a passing star, or perhaps by repeated encounters with the gravitational fields of the giant planets. The effects of comet impacts were highlighted in 1994, when Comet Shoemaker–Levy 9 broke into about twenty fragments after capture by Jupiter's gravitational field. After these kilometre-sized fragments impacted the planet, scars were long visible in Jupiter's outer atmosphere. Like asteroids, comets occasionally impact the Earth. A desire for our long-term survival, as well as a hunger for pure knowledge, will drive our exploration of these distant objects.

The medium between the planets is home to the rather chaotic solar wind. Highly variable, with an average density of about 10 atomic nuclei per cubic centimetre and a speed outward from the Sun of a few hundred kilometres per second, the solar wind begins to encounter the interstellar medium in the heliopause, located about 100 AU from the Sun. During the twentieth century, only the probes Pioneer 10/11 and Voyager 1/2 have been launched with trajectories allowing them to escape the heliosphere ('solar' space). When the still-active Voyagers cross the heliopause boundary in the first decades of the twenty-first century, they will become our first true emissaries to the Galaxy.

2.8 BIBLIOGRAPHY

Bakes, E., 'The Science Case for In-Situ Sampling of Kuiper Belt Objects', presented at STAIF 2000 Conference, University of New Mexico, Albuquerque, NM, January 30–February 3, 2000.

Brandt, J. C. and Hodge, P. W., *Solar System Astrophysics*, McGraw-Hill, NY (1964).

Butler, D. M., Newman, M. J. and Talbot, R. J., Jr., 'Interstellar Cloud Material: Contribution to Planetary Atmospheres', *Science*, **201**, 522–524 (1978).

Chaisson, E. and McMillan, S., *Astronomy Today*, 2nd ed., Prentice Hall, NJ (1996).

Chapman, C. R., 'Probing Europa's Third Dimension', *Science*, **283**, 338–339 (1999).

Davies, R. E., Delluva, A. M. and Koch, R. H., 'No Valid Evidence Exists for Interstellar proteins, Bacteria, etc.', in *The Search for Extraterrestrial Life: Recent Developments*, ed. M. D. Papagiannis, D. Reidel, Boston, MA (1985), pp. 165–169.

Dole, S. H., *Habitable Planets for Man*, Blaisdell, New York (1964).

Dole, S. H. and Asimov, I., *Planets for Man*, Random House, New York (1964).

Goody, R. M. and Walker, J. C. G., *Atmospheres*, Prentice Hall, Englewood Cliffs, NJ (1972).

Hansson, A., *Mars and the Development of Life*, 2nd edn., Wiley–Praxis, Chichester and New York (1997).
Heiles, C., 'The Interstellar Magnetic Field', *Annual Reviews of Astronomy and Astrophysics*, **14**, Annual Reviews Inc., Palo Alto, CA (1976), pp. 1–22.
Jastrow, R. and Rasool, S. I., 'Planetary atmospheres', in *Introduction to Space Science*, ed. W. N. Hess, Gordon and Breach, Philadelphia, PA (1965), Chapter 18.
Kasting, J. F., Whitmire, D. P. and Reynolds, R. T., 'Habitable Zones Around Main Sequence Stars', *Icarus*, **101**, 108–128 (1993).
Ya Kondratyev, K., *Radiation in the Atmosphere*, Academic, New York (1969).
Larson, R. B., 'Processes in Collapsing Interstellar Clouds', *Annual Reviews of Astronomy and Astrophysics*, **141**, Annual Reviews Inc., Palo Alto, CA (1973), pp. 219–238.
Wickramsinghe, N. C., Hoyle, F., Al-Mufti, S. and Wallis, D. H., 'Infrared Signatures of Prebiology – or Biology', in *Astronomical and Biochemical Origins and the Search for Life in the Universe*, ed. C. B. Cosmovici, S. Bowyer and D. Werthimer, Editrice Compositori, Bologna, Italy (1997), pp. 61–76.
Wickramsinghe, N. C. and Hoyle, F., 'Infrared Evidence for Panspermia: an Update', *Astrophysics and Space Science*, **259**, 385–401 (1998).
Wiegert, P. A. and Holman, M. J., 'The Stability of Planets in the Alpha Centauri System', *Astronomical Journal*, **113**, 1445–1450 (1997).

3

Tomorrow's targets

'Tis now the very witching time of night, when churchyards yawn,
And hell itself breathes out contagion to this world.

William Shakespeare, *Hamlet* (c.1602)

As the twenty-first century dawns, humanity is beginning to recover from its night-fear. Very soon we will challenge the cold and barren wastes that extend endlessly between the stars. It may be centuries yet before brave bands of human explorers enter this forbidding realm; but our robots will proceed us.

Most of us – scientists and laypeople alike – think of the space between the stars as a fairly uniform near-vacuum. But it is helpful for mission planners to divide the void into a number of separate regions, each with their own challenges and opportunities.

Perhaps the first engineer to develop a classification scheme for the zones of extrasolar space was Kraft Ehricke, in 1971. According to Ehricke, we can divide extrasolar space into a number of concentric zones centred on the Sun.

First is the solar magnetosphere, which we might today call the heliosphere. This is the region in which the motions of plasma streams are defined primarily by the Sun's magnetic field. The heliosphere extends perhaps 100 AU from the Sun. With Pioneers 10/11 and Voyagers 1/2 (which were ejected from the Solar System using giant-planet gravity assists), humanity has begun the exploration of the outer reaches of this zone. Perhaps the most interesting objects within the far heliosphere are the giant cometoids of the Kuiper Belt.

Next is the circumsolar zone, the transition region between interplanetary and interstellar influences. This region may extend a few thousand AU from the Sun and contains the Sun's gravity focus of 550 AU and nearer members of the Oort Cloud. To explore this region within a human lifetime requires Solar System exit velocities about ten times greater than those of Pioneer 10/11 and Voyager 1/2, and the development of various advanced propulsion options currently on the drawing board – the nuclear-electric drive, solar sails, and the solar-thermal drive. (The nuclear-electric drive and solar sail are considered in greater detail in the following chapters. The solar-thermal drive uses focused sunlight to heat hydrogen propellant to an exhaust velocity greater than $10\,\mathrm{km\,s^{-1}}$ – twice that of the Space Shuttle's main engines. See Salkeld *et al.* (1978) for an introduction to this propulsion system, and Maise *et al.* (1999) for its application to early extrasolar probes.) Robotic exploration of the circumsolar zone might commence within a few decades.

Finally, we come to the solar gravisphere, in which we expect to find distant members of the Oort Cloud bound gravitationally to the Sun and orbiting our star every 10,000–100,000 years. This region extends perhaps 60,000 AU (1 light year) from the Sun. A probe to 10,000 AU in a human lifetime requires another factor of 10 increase in spacecraft velocity. Hyperthin, aerogel or space-manufactured solar sails and nuclear-pulse propulsion are the best current candidates for missions to the solar gravisphere. We might sent our first robots towards this zone late in the twenty-first century. Since these craft could reach the nearest star system (α Centauri) in about 1,000 years, these late twenty-first century robots might be considered as humanity's first true starships.

3.1 THE TAU MISSION: AN EARLY NASA/JPL EXTRASOLAR MISSION STUDY

The first comprehensive study of an early twenty-first century extrasolar probe concept was published by L. D. Jaffe and other members of a NASA/Jet Propulsion Laboratory (JPL) study team in 1980. The mission goal for TAU was to propel a scientific payload in the year 2000 on an extrasolar trajectory such that the spacecraft traversed 1,000 AU during a 50-year flight time. The average velocity of the TAU probe relative to the Sun would be about $100\,\mathrm{km\,s^{-1}}$. About 10,000 years would be required for the TAU probe to reach the nearest extrasolar star system, if it were moving in the appropriate direction.

Many propulsive options were considered for TAU. These included direct launch from Earth; Jupiter gravity assist and powered flyby; solar and laser sailing; solar–electric and laser–electric propulsion; fusion; and antimatter. It was concluded that in the 2000 AD time-frame, only two propulsive options existed for the TAU mission: nuclear–electric propulsion (NEP), and the ultralight solar sail unfurled as close to the Sun as possible.

Primary TAU mission objectives included determination of heliopause and interstellar-medium characteristics, accurate measurements of stellar and galactic distances using long-baseline astrometry (possibly with two or more TAU probes moving along different trajectories), examination of cosmic rays excluded by the heliosphere, and determination of large-scale solar system characteristics from afar. Secondary mission objectives included close observation of Pluto (if a TAU probe were directed to pass near that planet), extragalactic observations, and evaluation of the possibilities of observing other solar systems from spacecraft.

Candidate scientific instruments for TAU included magnetometers and electric-field meters, spectrometers and radiometers, radio-astronomy detectors and optical cameras. Excluding power and propulsion, the estimated mass of the TAU probe was 1,200 kg. Estimated data-transmission rates were in the region of 10 kilobits per second. Astrometric photographs of distant galactic objects could be transferred to Earth at the rate of several images per day.

Exercise 3.1 Because of the speed-of-light limitation, an extrasolar or interstellar probe requires a high degree of intelligence and autonomy. Estimate the time required for a signal radioed from the Earth to reach a probe at 100, 500 and 1,000 AU from the Sun.

Sadly, the TAU study did not lead directly to an extrasolar mission. It does, however, serve as a valuable baseline study for contemporary extrasolar mission plans.

3.2 SETISAIL AND ASTROSAIL: PROPOSED PROBES TO THE SUN'S GRAVITY FOCUS

In the late-1980s, Alenia Spazio – an aerospace company in Turin, Italy – proposed Quasat, an inflatable Earth-orbiting radio telescope. Although Quasat never flew, Claudio Maccone (a theoretical physicist associated with Alenia) has coordinated efforts to investigate extrasolar applications of Quasat inflatable technology. Starting in 1992, regular meetings have taken place in northern Italy to further investigate these ideas.

Rather than particles, fields and astrometric measurements, Quasat-derived 'focal' probes have been proposed to utilise the Sun's gravity focus at 550 AU for radio-astronomical purposes. The potential of the Sun's gravity focus for interstellar observations and communications was first considered by V. Eshelman in 1979, even though gravity lenses have been understood for decades as a consequence of general relativity theory, and many have been discovered in intergalactic space.

Figure 3.1 (which is not to scale) schematically presents the focusing of electromagnetic (EM) waves by the Sun's gravitational field. Electromagnetic waves are emitted by an (occulted) object on the other side of the Sun from a spacecraft, which is located at least 550 AU from the Sun. Beyond 550 AU, the EM radiation from the occulted object is amplified by a factor of about 10^8. Unlike optical lenses, in which the light diverges after the focus, the gravity-focused radiation remains along the focal axis for solar separations greater than 550 AU. The 'spot radius' (distance from

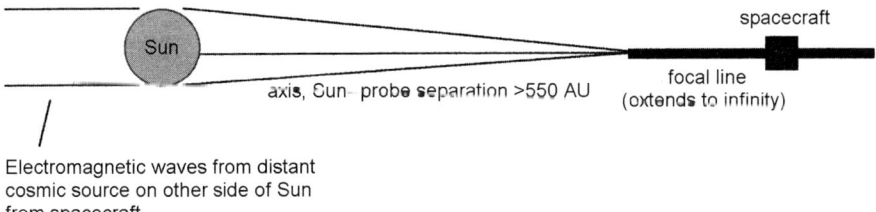

Fig. 3.1. The Sun as a gravitational lens.

the centre line of the image at which the image intensity gain falls by a factor of 4) has been calculated by Eshelman to be about 11 km for a Sun–spacecraft separation of 2,200 AU. As demonstrated by Kraus, the off-axis gain decreases with the inverse square root of the off-axis distance.

The distance between the Sun and the minimum focal distance of the Sun's gravitational lens can be calculated using the following simple result from general relativity:

$$D_{\text{solar-focal}} = \frac{R_{\text{Sun}}^2 c^2}{4GM_{\text{Sun}}} \quad (3.1)$$

where R_{Sun} is the Sun's radius, c is the speed of light, G is the gravitational constant, and M_{Sun} is the Sun's mass.

Exercise 3.2 By substitution into equation (3.1), verify that the distance between the Sun and the near-focus of the solar gravitational lens is about 550 AU or 3.2 light days.

It should not be assumed that deconvoluting data received from a solar gravity focus probe will be a simple matter. Because deflection is greater for light rays passing farther from the Sun's centre, the probe will detect a ring of light called an Einstein Ring from a point cosmic EM source occulted by the Sun. As discussed by Kaler, extended objects occulted by the Sun observed at the Sun's gravity focus might produce multiple images like the Einstein Cross which has been detected for the case of quasers occulting more distant galaxies.

In this analysis of Quasat-derived inflatable-solar-sail probes to the Sun's gravitational focus, Matloff considered a number of options for probe design, mission and performance. An inflatable solar sail, as shown in Figure 3.2, consists of an inflatable plastic structure and a reflective metallic layer facing the Sun. Thrust is obtained by the radiation pressure of sunlight reflected from the reflective layer (as discussed in the following chapters). Waste heat is radiated from the two plastic layers. It is usually assumed that the mass of the gas required to inflate the sail in negligible.

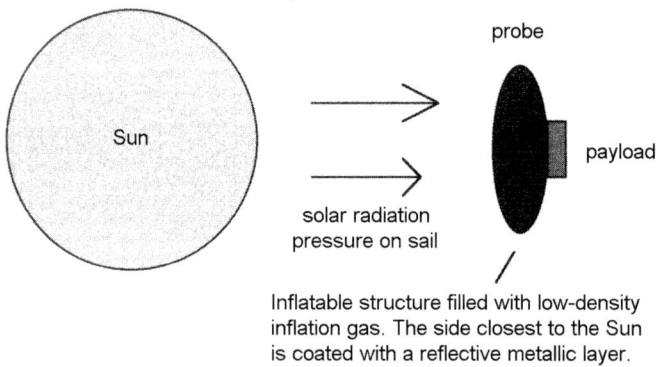

Fig. 3.2. An inflatable solar-sail probe to the Sun's gravity focus.

Spacecraft performance is enhanced by high reflectivity, high plastic operating temperature (which allows a closer approach to the Sun) and low mass for sail and payload. Matloff's baseline inflatable sail had an area of 10,000 square metres, an operating temperature of 488 K, a thickness of 2 µm, a plastic-layer specific gravity of 0.9, an aluminium reflective layer with a thickness of 0.1 µm and a reflectivity of 0.9 and a payload of 10–20 kg. The total mass of the spacecraft, excluding an upper stage required to escape low Earth orbit (LEO), is less than 100 kg – well within the range of small launch vehicles. The time required to reach the Sun's gravity focus at 550 AU is about 60 years, which could be reduced by Jupiter gravity assist.

If radioactive-isotope thermal generators remain environmentally acceptable for deep space missions and can be reduced in mass, these represent a possible power source for focal probes; otherwise, low-thickness, high-efficiency solar cells must coat much of the spacecraft surface, or onboard power must be obtained at the expense of spacecraft motion through the interstellar magnetic field.

One advantage of a sail-powered focal probe with a variable shape is that the sail could remain unfurled or inflated beyond the Sun's near gravity focus to serve as an EM collecting surface. But the challenges of performing sail reconfiguration within a 10–20 kg payload mass budget are formidable.

Many astrophysical targets exist for study by focal probes. These include globular clusters and external galaxies, the million-solar-mass black hole suspected of lurking at our Galaxy's centre, and recently discovered planets orbiting nearby stars. Heidmann and Maccone have suggested that two classes of focal probe be launched – one to study objects of astrophysical interest (ASTROsail) and one to study suspected artificial EM signals from other civilizations (SETIsail).

Maccone has suggested an interesting trajectory option for focal probes that could increase payload or decrease flight time – the double jovian-flyby. In such a mission, the spacecraft first visits Jupiter, where the planet's gravity field directs it towards a perihelion closer to the Sun than Mercury in a parabolic or slightly hyperbolic orbit. The sail is used at perihlion to accelerate to interstellar velocities and later redirect the craft towards a second Jupiter flyby. Further acceleration is effected by the second Jupiter gravity assist. This suggested trajectory option deserves further analysis.

3.3 THE AURORA PROJECT: A SAIL TO THE HELIOPAUSE

Recognising that a focal probe might be a challenging venture for humanity's first dedicated interstellar spacecraft, some members of the mostly European team involved with the focal proposal turned their attention to the somewhat less ambitious Aurora project. A meeting of the International Academy of Astronautics in Turin, Italy, during June 1996 was largely devoted to the presentation of preliminary Aurora results.

As discussed by Giovanni Vulpetti, a mass/area ratio for a contemporary solar sail spacecraft is 0.0015–0.0025 kg/m^2. Proper trajectory design and a close solar pass can result in a Solar System exit velocity of at least 12 AU/year – about three times greater than that of the Voyager probes. The total spacecraft mass, including payload, structure, and flat-sheet sail, is about 150 kg.

Gabriele Mocci considered the spacecraft–Earth communication link, and concluded that a mission to 50–100 AU from the Sun is feasible using current-technology communication hardware carried as payload onboard the sailcraft. Preliminary structural analysis of this sail to the heliopause was reported in a paper by G. Genta and E. Brusa. A thin-film 250-m square sail supported by booms and struts constructed using carbon reinforced plastics is a structurally feasible sail layout.

In a paper by S. Santoli and S. Scaglione (further discussed and listed in the references for the next chapter), an innovative approach to the reduction of sail areal mass thickness was discussed. A bilayer sail consisting of an aluminium reflective coating attached to a plastic substrate would be launched from the Earth and unfurled in space. The plastic substrate would be chosen to evaporate under exposure to solar ultraviolet radiation. Although innovative, we shall see that this approach may be unnecessary.

3.4 THE NASA INTERSTELLAR INITIATIVE

In response to the interest in interstellar travel expressed by NASA Administrator Dan Goldin, a serious effort is underway at various NASA centres directed towards the achievement of an extrasolar/interstellar robotic capability in the twenty-first century. The NASA interstellar initiative is a response to Goldin's challenge, and is described in a number of papers, including Les Johnson's and Stephanie Leifer's contributions to the NASA Interstellar Probe definition study.

The first mission would be a sail past the heliopause, to be launched around 2010. This would soon be followed by a craft designed to rendezvous with a Kuiper Belt object. Later in the century, true interstellar robots might be directed to visit destinations in the Oort Cloud, as far as 10,000 AU from the Sun.

3.5 THE NASA HELIOPAUSE SAIL

As reported by Charles Garner *et al.* in their 1999 AIAA paper, a number of different materials have been considered for the solar sail of the heliopause mission. Most exciting is a high-temperature, thick, strong and porous fabric developed by Energy Science Laboratories Inc. (ESLI) in San Diego, California. This material – a carbon microtruss – consists of a three-dimensional mesh of interconnected carbon microfibres. Many properties of the new material are described by Timothy Knowles *et al.* of ESLI. Achievable properties most applicable to solar-sail application of the new carbon microtruss are listed in Table 3.1, with the assumption of a 90% reflective aluminium layer deposited on top of the microtruss. Further research indicated that high mesh reflectivity might render an aluminium reflective layer unnecessary.

Table 3.1. Properties of a carbon microtruss sail coated with an aluminium reflective layer

Minimum carbon microtruss layer areal mass thickness	2.00×10^{-4} kg m^2
Minimum aluminium-reflective layer areal mass thickness	6.75×10^{-5} kg m^2
Aluminium layer fractional reflectivity to sunlight	0.9
Carbon microtruss layer emissivity range	0.4–0.9
Sail material operational temperature range	70–2,000 K
Sail tensile strength (measured at 300 and 525 K)	2,205 megaPascal

During his tenure as a 1999 Summer Faculty Fellow at the NASA Marshall Space Flight Center (MSFC) in Huntsville, Alabama, the author evaluated extrasolar mission possibilities using the ESLI sail material, using analytical techniques described in the next chapter. Consider, for example, the case of a 27-kg science payload; a non-science payload (command, control, propulsion) of 86 kg; and a mass for sail interface, contingency, and fuel of 119 kg. The total sail mass is 1,063 kg, for a 1-km radius sail constructed from the material described in Table 3.1. If the sail is fully unfurled at 0.2 AU from the Sun's centre (well within this material's thermal and tensile capabilities) at the perihelion of an initially parabolic sail orbit, and is oriented normal to the Sun, if exits the Solar System at about 200 km s^{-1}.

Such a craft could reach 500 AU from the Sun in about 12 years, and 1,000 AU in about 24 years. Even a much smaller sail, or an inflatable constructed using this material, could explore the heliopause, reach the Sun's gravity focus and accomplish a TAU mission well within a human lifetime.

3.6 THE NASA KUIPER BELT EXPLORER

As discussed by Alessandro Morbidelli, the Kuiper Belt consists of tens of thousands of icy objects larger than 100 km, in eccentric high-inclination orbits 35–45 AU from the Sun. William Ward and Joseph Hahn have pointed out an orbital interaction between Neptune and the Kuiper Belt objects.

Perhaps such giant-planet perturbations converts Kuiper Belt objects (KBOs) into short-period comets that occasionally collide with the Earth, resulting in mass extinctions of terrestrial life. Juan Oro and Cristiano Cosmovici have recently reviewed the evidence that cometary impacts seeded the primeval Earth with organic precursors to life. Although less likely, we cannot completely rule out biospheres within some of the Solar System's small, icy bodies.

A valuable deep-space goal for the NASA interstellar initiative is rendezvous missions with KBOs that deposit experiment packages on the surface of the KBOs to measure geological, chemical and physical properties. Such missions would tell us a great deal about the origins of terrestrial life, might help us plan Earth-defence projects to deflect or destroy future cometary threats, and might also inform us about extraterrestrial life.

32 Tomorrow's targets   [Ch. 3

We might consider an undecelerated sail-launched mission that deposits instrument packages encased in penetrator probes; but such probes would constitute kinetic weapons when impacting a low-tensile strength KBO.

Exercise 3.3 Consider a 10-kg surface-penetrator probe released from a KBO-flyby probe at a velocity relative to the KBO of $50\,\text{km}\,\text{s}^{-1}$. Calculate the penetrator's kinetic energy relative to the KBO at the moment of impact. If the escape velocity of the KBO is $1\,\text{km}\,\text{s}^{-1}$, how much material will be blown off of the KBO if 50% of the penetrator's kinetic energy is converted into impact-debris kinetic energy.

Although dramatic from a special effects point of view, such an explosive impact would yield little scientific data. Some form of deceleration in the Kuiper Belt seems to be essential.

Since no known solar approach will efficiently decelerate a spacecraft 40 AU or so from the Sun, some sort of nuclear drive seems necessary to perform the KBO rendezvous mission. One approach might be to launch the KBO object rendezvous probe from Earth, apply a Jupiter gravity assist manoeuvre to exit the Solar System at the speed of Voyager 1 (about 3.5 AU per year relative to the Sun) and decelerate to rendezvous with the KBO using nuclear–electric propulsion (NEP). Described in following chapters, the NEP uses energy released from an onboard nuclear fission reactor to accelerate ionised fuel particles to exhaust velocities as high as $100\,\text{km}\,\text{s}^{-1}$.

Exercise 3.4 A probe is launched from Earth to rendezvous with a KBO at 40 AU from the Sun, at zero velocity relative to the KBO. Estimate the mission duration for the case of low- and high-energy trans-Jupiter trajectories respectively requiring five and two years, an average post-Jupiter cruise velocity of 3.5 AU per year relative to the Sun, and a one-year deceleration interval.

Even with high-performance NEP engines (such as the design recently suggested by Roger Lenard and Ronald Lipinski), the mission time for a KBO rendezvous mission will be measured in decades. As well as contending with the environmental issues involved with flying a space-qualified reactor, mission designers must shield payloads from the nuclear radiation. While the mass of the heliopause sail will be measured in hundreds of kilogrammes, the KBO rendezvous probe will almost certainly have a mass of a few thousand kilogrammes. But because of the scientific payoff of such a mission and the desire of many people to ultimately protect the Earth from cometary impacts, there is a good chance that a KBO rendezvous mission will be flown in the first decades of the twenty-first century.

3.7 A PROBE TO THE OORT CLOUD

If we take the ESLI carbon-microtruss sail material considered for the heliopause sail to its thermal limits and unfurl the sail at a closer perihelion distance (0.02–0.03 AU) from the Sun's centre, the craft will be capable of performing a flight to 10,000 AU – well within the Oort Cloud – within a human lifetime. An exploratory mission of the near-Sun radiation and field environment (such as that suggested by R. L. McNutt Jr. *et al.*) is a necessary prelude to true interstellar solar-sail missions utilising very close perihelion passes.

The Oort Cloud probe could be launched after 2020, its design based upon the results of a near-Sun explorer. Such an advanced solar-sail robot (which would require about 1,000 years to reach α Centauri) might be the best we can achieve without investing in the space infrastructure necessary to develop laser-pushed sails, fusion rockets or antimatter propulsion. That is, unless an unexpected and unpredictable breakthrough occurs to drastically reduce interstellar transit times. The possibility of such a breakthrough is taken seriously by NASA mission planners, and is discussed in later chapters.

3.8 BIBLIOGRAPHY

Ehricke, K. A., 'The Ultraplanetary Probe', AAS-71-164.

Einstein, A., 'Lens-like Action of a Star by the Deviation of Light in the Gravitational Field', Science, **84**, 506–507 (1936).

Eshelman, V., 'Gravitational Lens of the Sun: its Potential for Observation and Communication over Interstellar Distances', *Science*, **205**, 1133–1135 (1979).

Garner, G., Diedrich, B. and Leipold, M., 'A Summary of Solar Sail Technology Developments and Proposed Demonstration Missions', AIAA-99-2607.

Genta, G. and Brusa, E., 'Project Aurora: Preliminary Structural Definition of the Spacecraft', in *Missions to the Outer Solar System and Beyond, 1st IAA Symposium on Realistic Near-Term Scientific Space Missions*, ed. G. Genta, Levrotta & Bella, Turin, Italy (1996), pp 25–36.

Heidmann, J. and Maccone, C., 'ASTROsail and SETIsail: Two Extrasolar System Missions to the Sun's Gravitational Focus', *Acta Astronautica*, **37**, 409–410 (1994).

Jaffe, L. D., Ivie, C., Lewis, J. C., Lipes, R., Norton, H. N., Sterns, J. W., Stimpson, L. D. and Weissman, P., 'An interstellar Precursor Mission', *Journal of the British Interplanetary Society*, **33**, 3–26 (1980). (Included in this text are the results of C. Uphoff's unpublished consideration of extrasolar travel by solar sail.)

Johnson, L. and Leifer, S., 'Propulsion Options for Interstellar Exploration', AIAA-2000-3334.

Kaler, J. B., *Astronomy!*, Addison Wesley, Reading, MA (1997).

Knowles, T. R., Oldson, J. C. and Liew, P., 'Microtruss composite Sails', presented at NASA/JPL/MSFC/AIAA Annual Tenth Advanced Space Propulsion Workshop, Huntsvill, AL, April 5–8, 1999.

Kraus, J. D., *Radio Astronomy*, Cygnus-Quasar Books, Powell, OH (1986).

Lenard, R. X. and Lipinski, R. J., 'Architecture for Fission-Powered Propulsion to Nearby Stars', presented at NASA/JPL/MSFC/AIAA Annual Tenth Advanced Space Propulsion Workshop, Huntsville, AL, April 5–8, 1999.

Maccone, C., 'The Quasat Satellite and its SETI Applications', in *Bioastronomy: The Next Steps*, ed. G. Marx, Kluwer Academic Publishers, Norwell, MA (1988).

Maise, G., Powell, J. and Paniagua, J., 'SunBurn: A New Concept Enabling Ultra High Spacecraft Velocities for Extra Solar System Exploration', IAA-99-IAA.4.1.07.

Matloff, G. L., 'Solar Sailing for Radio Astronomy and SETI: An Extrasolar Mission to 550 AU', *Journal of the British Interplanetary Society*, **47**, 476–484 (1994).

McNutt Jr., R. L., Krimigis, S. M., Cheng, A. F., Gold, R. E., Farquhar, W., Roelof, E. C., Coughlin, T. B., Santo, A., Bokulic, R. S., Reynolds, E. L., Williams, B. D. and Wiley, C. E., 'Mission to the Sun: the Solar Pioneer', *Acta Astronautica*, **35**, Supplement, 247–255 (1995).

Morbidelli, A., 'New Insights on the Kuiper Belt', *Science*, **280**, 2071–2073 (1998).

Mocci, G., 'The Aurora Project: Analysis of Impacts of the Communication Requirements on Mission Proficiency', in *Missions to the Outer Solar System and Beyond, 1st IAA Symposium on Realistic Near-Term Scientific Space Missions*, ed. G. Genta, Levrotto & Bella, Turin, Italy (1996), pp. 17–24.

NASA, 'Interstellar Probe Science and Technology Definition Team Meeting #1, February 15–17, 1999, and Meeting #2, March 29–31, 1999', NASA/JPL, Pasadena, CA. Also see P. C. Liewer, R. A. Mewaldt, J. A. Ayon and R. A. Wallace, 'NASA's Interstellar Probe Mission', and C. E. Garner, W. A. Layman, S. A. Gavit and T. Knowles, 'A Solar Sail Design for a Mission to the Near Interstellar Medium', both presented at STAIF 2000 Conference, University of New Mexico, Albuquerque, NM, January 30–February 3, 2000.

Oro, J. and Cosmovici, C. B., 'Comets and Life on the Primitive Earth', in *Astronomical and Biochemical Origins and the Search for Life in the Universe*, ed. C. B. Cosmovici, S. Bowyer and D. Werthimer, Editrice Compositori, Bologna, Italy (1997), pp. 97–120.

Salkeld, R., Patterson, D. W. and Grey, J., *Space Transportation Systems*, American Institute of Aeronautics and Astronautics, Reston, VA (1978).

Vulpetti, G., 'The Aurora Project: Flight Design of a Technology Demonstration Mission', in *Missions to the Outer Solar System and Beyond, 1st IAA Symposium on Realistic Near-Term Scientific Space Missions*, ed. G. Genta, Levrotto & Bella, Turin, Italy (1996), pp. 1–16.

Ward, W. R. and Hahn, J. M., 'Neptune's Eccentricity and the Nature of the Kuiper Belt', *Science*, **280**, 2104–2106 (1998).

4

Space propulsion today

Allons! The inducements shall be greater,
We will sail pathless and wild seas,
We will go where winds blow, waves dash
And the Yankee Clipper speeds by under full sail.

Walt Whitman, *Song of the Open Road*

Since the dawn of the Space Age, hundreds of humans have entered this strange new realm in modern-day equivalents of the *Yankee Clipper*. Twenty-four (three of them twice) have orbited our Moon or landed upon it and viewed the Earth as a precious blue-green orb suspended in the inky blackness of the void. Our robot emissaries have tested the soils of the Moon, Mars and Venus, and have flown by all Solar System planets except frozen Pluto. The robotic exploration of local asteroids and comets and the satellites of Solar System planets continues, and four small craft – the intrepid Pioneer 10/11 and Voyager 1/2 – have become humanity's first Galactic emissaries.

Most of this progress is due to application of an ancient Chinese invention – the chemical rocket. Gravity assist flybys of Solar System planets and satellites has also found application. The first solar–electric (ion drive) propelled rockets have been tested in space, and experimental solar-photon sails have been unfurled in orbit.

Admittedly, projection of existing technologies does not allow the eventual realisation of science fiction's dream of rapid spaceflight to stars. However, the realm of the comets and nearby interstellar space are within reach. And using projections of current solar-photon sail technology, travel to the nearest stars beyond the Sun may require 'only' a millennium-duration journey.

4.1 ROCKET HISTORY AND THEORY

The rocket has existed for a long time. The first reaction device may have been the tethered, steam-propelled 'wooden dove' of Archytas, which was constructed in what is now southern Italy, in about 360 BC. In about 50 BC, Hero of Alexandria constructed his famous steam-powered aeolipile, in which pressurised steam was vented from a tethered hollow sphere, causing rotation in a manner analogous to a lawn sprinkler.

But these early models served mainly to demonstrate the action–reaction principle. The first 'operational' flight rockets made their debut in China by about 1040 AD. They were propelled by 'black powder', and were applied as crude fireworks and artillery. Crude rockets launched from crossbows were apparently used to defeat a Mongolian calvary attack upon Peking in 1230 AD.

Some time during the late Middle Ages, the first attempt at human spaceflight may have occurred in China. As the legend tells us, a somewhat world-weary official named Wan Hu had two kites and 47 rockets attached to his chair. When his assistants lit the fuse, Mr Hu and his apparatus vanished in an enormous flash. Perhaps he became the first human spacefarer; more likely, he was blown to smithereens.

Brought to Europe via Italian merchant adventurers in about 1260 AD, the rocket concept was combined with an Arabic invention – gunpowder – to produce the first major firework displays. In the late eighteenth century, William Congreve, in England, was experimenting with rockets that would ultimately be used for naval bombardment and as a life-saving tool.

With the military application of rockets throughout Europe well established before the nineteenth century, it is surprising that early science fiction writers such as H. G. Wells resorted to exotic devices such as 'space guns' instead of interplanetary rockets in their classic tales of interplanetary exploration and warfare. The first serious attempts to develop the rocket principle for space application awaited early twentieth century pioneers such as Konstantin Tsiolkovsky in Russia, Herman Oberth in Germany and Robert Goddard in the USA.

Tsiolkovsky is best known for his theoretical development of rocket mechanics, Goddard flew the first liquid-fuelled chemical rockets, and Oberth's work culminated in the V2. The V2 (Vengance 2) was used by Nazi rocketeers to bombard London during World War II, and was the first rocket capable of travelling above Earth's atmosphere.

During the post-war era, American and Russian engineers were aided by captured German experts in developing improved versions of the liquid-fuelled V2. By 1957 these were capable of placing small satellites in Earth orbit. More powerful rockets were applied throughout the 1960s to first place human-occupied spacecraft in orbit and ultimately to visit the surface of the Moon.

We may consider the operation of a rocket using the diagram in Figure 4.1. Before a small fuel mass, dM_f, is released, a rocket of mass M_S is moving with a velocity of V_s relative to a reference frame on the surface of the Earth. The fuel is expelled with an exhaust velocity V_e relative to the spacecraft. After the fuel has been expelled, the rocket moves at $V_s + dV_s$ and the fuel moves at $V_s - V_e$, relative to the reference frame.

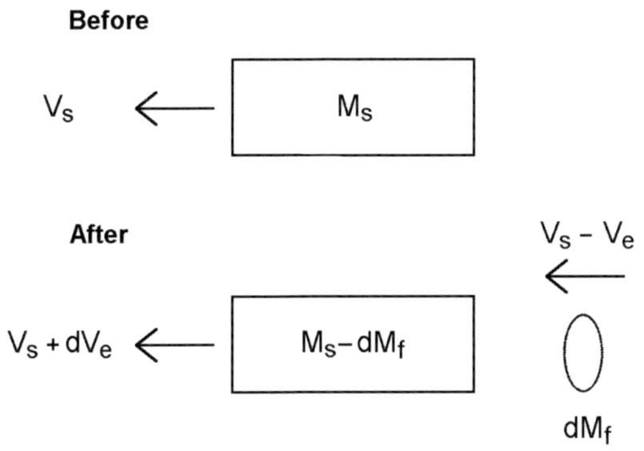

Before

$V_s \leftarrow \boxed{M_s}$

After

$V_s + dV_e \leftarrow \boxed{M_s - dM_f}$ $\overset{V_s - V_e}{\leftarrow}$ $\bigcirc$
dM_f

Fig. 4.1. The rocket principle.

Equating linear momentum of the system before and after the tiny fuel mass is expelled,

$$M_s V_s = (V_s + dV_s)(M_s - dM_f) + dM_f(V_s - V_e) \quad (4.1)$$

Simplifying and rearranging,

$$M_s dV_s = V_e dM_f \quad (4.2)$$

Equating the ship's mass change of $-dM_s$ with the expelled fuel mass of dM_f, equation (4.2) can be rearranged for integration:

$$\int_0^{V_f} \frac{dV_s}{V_e} = -\int_{M_0+M_f}^{M_0} \frac{dM_s}{M_s} \quad (4.3)$$

where the rocket has zero initial velocity relative to the reference frame and a velocity V_f after all the fuel is exhausted, M_0 is the empty (fuelless) rocket mass, and M_f is the total mass of fuel expelled.

Defining the mass ratio (MR) as the ratio of $(M_0 + M_f)$ to M_0, equation (4.3) can be integrated to obtain

$$MR = e^{V_s/V_e} \quad (4.4)$$

which is the standard form of the non-relativistic rocket equation.

If we next define rocket thrust F as the product of exhaust velocity and (dM_f/dt), the rate at which fuel mass is expelled with time t, and the rate at which rocket weight W changes with time (dW/dt) as the product of gravitational acceleration near Earth's surface, g, and (dM_f/dt), we find that

$$\frac{F}{dW/dt} = \frac{V_e dM_f/dt}{g dM_f/dt} = \frac{V_e}{g} \quad (4.5)$$

where the ratio V_e/g is the specific impulse (I_{sp}), measured in units of seconds.

The best contemporary chemical rockets – such as the liquid hydrogen/oxygen main engines of the Space Shuttle – have a specific impulse less than about 500 seconds, or an exhaust velocity less than $5 \, \text{km s}^{-1}$. Application of equation (4.4) reveals that a rocket with $I_{sp} = 500$ seconds requires a mass ratio of about 5 to achieve a velocity of $8\,\mathrm{km\,s^{-1}}$ – the velocity necessary for a low Earth circular orbit. If the useful payload comprises 10% of the unfuelled spacecraft mass on the launching pad, only 2% of the total mass on the pad (including fuel) is payload.

The low payload fraction is the reason why most rockets are staged. Design problems are minimised by not carrying expended engines and empty fuel tanks all the way to orbit. Consider, for example, the Space Shuttle. The empty orbiter mass is about 75,000 kg, the empty external-tank mass is 35,500 kg and the unfuelled mass of the solid rocket boosters is about 84,100 kg. The maximum payload capacity is 29,500 kg, and the mass of the entire vehicle fuelled for flight is 2 million kg. Even with staging (the solid boosters and external tank are not carried all the way to orbit), the mass ratio is about 8.9. Useful payload comprises less than 2% of the total mass on the launch pad.

Improvements in chemical rocket technology may increase, but will probably not double the specific impulse of future chemical rockets within the foreseeable future. Table 4.1 outlines specific impulses for a number of existing and projected chemical propellant combinations.

In addition to the proven propellant combinations listed in Table 4.1, Mallove and Matloff mention some unproven and very exotic combinations thay may one day be capable of very high performance. These include free radicals (H + H → H_2), which has a theoretical I_{sp} of 2,130 seconds and reactions of atoms in metastable atomic states, such as helium, which has a theoretical I_{sp} of 3,150 seconds.

Exercise 4.1 Validate all steps in the derivation of equation (4.4), and then calculate mass ratios for single-stage rockets for ascent to low Earth orbit for some of the propellant combinations in Table 4.1. Repeat this exercise for the case of direct ascent from the Earth's surface to an Earth-escape velocity of $11\,\mathrm{km\,s^{-1}}$.

Table 4.1. Specific impulses of some chemical propellants

Propellant/oxidiser combination	Specific impulse (seconds)
Kerosene/nitric acid	255
Kerosene/liquid oxygen	306
Hydrazine/chlorine pentafluorine	306
Hydrazine/liquid oxygen	306
Pentaborane/hydrazine	327
Liquid hydrogen/liquid oxygen (ideal)	528
Liquid hydrogen/liquid oxygen (Space Shuttle)	460
Liquid fluorine/liquid oxygen	530
Hydrogen/ozone	607
Lithium hydride (Li-H_2)/fluorine	703
Beryllium hydride (Be-H_2)/oxygen	705

Sources: Forward (1988), p. 130; Mallove & Matloff (1989), p. 43.

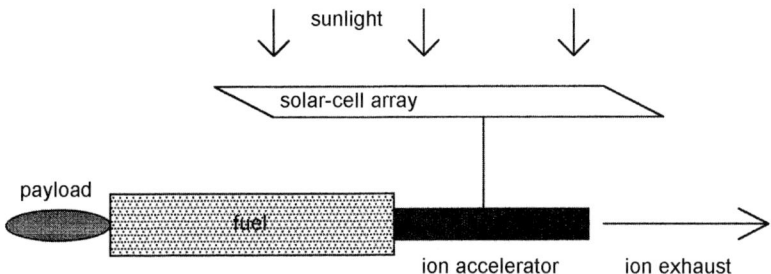

Fig. 4.2. The solar-electric drive.

4.2 THE SOLAR-ELECTRIC DRIVE

One method of obviating the specific-impulse limitations of conventional chemical rockets is the solar-electric or 'ion' drive. Long a favourite of science fiction authors, the solar-electric drive is the prime propulsion system for Deep Space 1, a robotic asteroid-flyby mission launched by NASA in 1998.

Operation of a typical solar-electric drive is presented in Figure 4.2. Sunlight falls on a solar cell array with an area of A_{sc}. These solar cells (which are here assumed to be oriented normal to the sunlight) convert incident sunlight into electrical energy with an efficiency of ε_{sc}. Sun-derived electricity is then used to ionise fuel atoms (usually argon, caesium or mercury) and accelerate the ionised fuel to exhaust velocity V_e in an ion accelerator. The electrical energy is converted into ion kinetic energy with an efficiency of ε_{ia}. The specific power, P_{sp}, of the thruster subsystem is defined as the power output (in kilowatts) divided by the thruster mass in kilogrammes. The thruster power, P_{th}, can be related to specific power and thruster mass, M_{th}, by the equation

$$P_{th} = 1000 P_{sp} M_{th} \text{ W} \qquad (4.6)$$

Thruster power can also be related to exhaust velocity, system efficiencies and solar array area:

$$P_{th} = \frac{1}{2}\frac{dM_f}{dt} V_e^2 = \varepsilon_{ia}\varepsilon_{sc} A_{sc} S_c \text{ W} \qquad (4.7)$$

In equation (4.7) the term S_c refers to the solar energy per second incident on the solar cell array ($S_c = 1,400/R_{au}^2$ W m^{-2}, where R_{au} is the distance to the Sun in Astronomical Units).

The performance of any solar-electric propelled spacecraft can be evaluated using equations (4.4), (4.6) and (4.7). Consider, for example, a fairly advanced electric propulsion craft with the following characteristics:

$P_{sp} = 0.2 \text{ kw/kg}$, $\varepsilon_{ia} = 0.6$, $\varepsilon_{sc} = 0.2$, and $A_{sc} = 1200 \text{ m}^2$, $V_e = 100 \text{ km s}^{-1}$.

The specific gravity of the solar array is assumed to be 2, and the array thickness is 10 μm.

At a distance of 1 AU from the Sun, the thruster power is calculated, from equation (4.7), as 2×10^5 W. Also from equation (4.7), $dM_f/dt = 4 \times 10^{-5}$ kg s^{-1}. The mass of the solar array is found, by multiplying its thickness, density and area, to be 24 kg. Applying equation (4.6) we see that the mass of the thruster is 1,000 kg.

If the thruster operates for one year, about 1,260 kg of ion fuel will be exhausted. If the unfuelled spacecraft mass is 1,500 kg, the mass ratio is 1.84. Applying equation (4.4) for this value of mass ratio and a 100 km s^{-1} exhaust velocity, we find that the spacecraft's velocity changes by 61 km s^{-1}. During the one-year thrusting period, the average spacecraft acceleration is 1.93×10^{-3} m s^2, or about 0.0002 g (Earth surface gravities).

Exercise 4.2 Check all calculations outlined above for this hypothetical solar-electric spacecraft operating near the Earth. Repeat the results for the same spacecraft operating near Jupiter (at about 5 AU from the Sun). Then repeat the calculations for a less advanced solar-electric craft with a specific impulse of 3,000 seconds and a specific power of 0.03 kW kg^{-1}, similar to the performance possible using Deep Space 1 technology.

During the year-long acceleration period, the spacecraft will of course not remain at 1 AU from the Sun, so these results are indicative of solar-electric drive performance rather than being accurate kinematics. Still, it is obvious that the solar-electric drive must be a highly reliable device, capable of maintaining fairly constant thrust during the month or year duration time intervals required for this low-acceleration, high exhaust-velocity engine to achieve its terminal velocity.

4.3 UNPOWERED PLANETARY GRAVITY ASSISTS

Gravity-assist manoeuvres are now essential tools in launching spacecraft to the outer Solar System and beyond. Humanity's first outer-planet and extrasolar probes, Pioneer 10/11 and Voyager 1/2, made ample use of unpowered giant-planet gravity assists to redirect their velocity vectors; the solar probe Ulysees applied a Jupiter gravity-assist to manoeuvre into an orbit passing over the Sun's poles; Mariner 10 used a Venus flyby to repeatedly visit the vicinity of Mercury; Jupiter-orbiter Galileo and Saturn-probe Cassini have used multiple flybys of Earth and Venus to reach the outer planets.

There are actually two types of gravity assist manoeuvres. These are: an unpowered planetary flyby, in which the spacecraft uses a planet's gravitational field to redirect its velocity vector relative to that planet; a close flyby of a large celestial body with a powered periapsis manoeuvre deep within the gravitational field 'well' of that celestial object. (To review gravity assist literature, see L. Ravenni (1997)).

Although missions to date have generally applied the unpowered approach, NASA engineer K. Nock has suggested that the highest Solar System exit velocities possible using conventional technology can be achieved using a close flyby of the Sun with a chemical rocket burn at perihelion.

Sec. 4.3]   Unpowered planetary gravity assists 41

Essentials of unpowered planetary flybys can be understood referring to Figure 4.3, which presents two extreme cases. In both cases, the planet (P) orbits the Sun (S) with a velocity of V_{planet}, the spaceprobe's velocity relative to the planet is $V_{srp,bef}$ before the encounter, and the spaceprobe's velocity relative to the planet is $V_{srp,aft}$ after the encounter. During an unpowered planetary encounter, the direction of the probe's velocity relative to the planet changes, but the magnitude of the velocity does not.

In Figure 4.3(a), the probe is in an initial retrograde solar orbit (moving in the opposite direction around the Sun from the planet). Its pre-encounter velocity relative to the Sun is $V_{srp,bef} - V_{planet}$. Since the probe's trajectory in this case has been

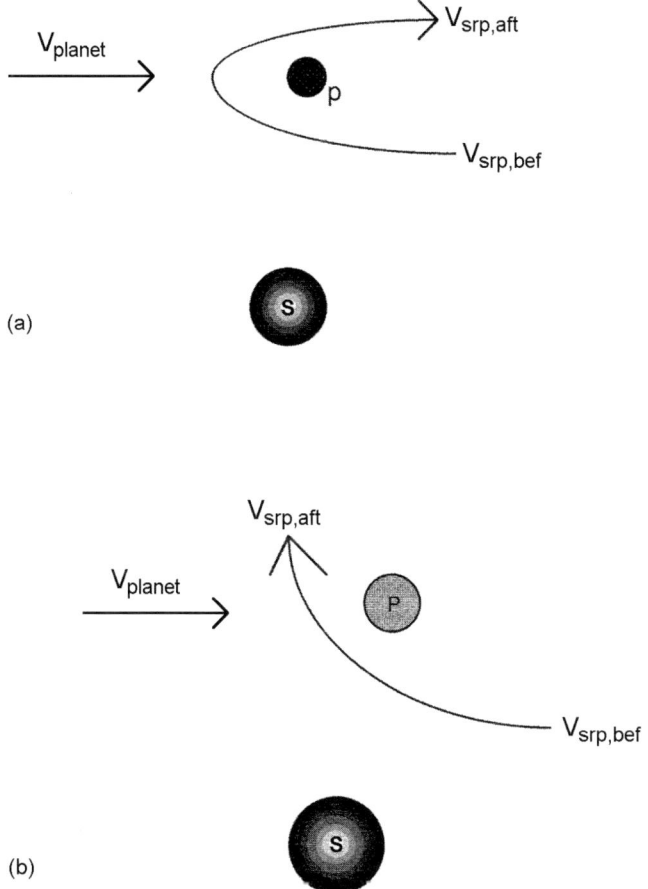

Fig. 4.3. Representation of an unpowered planetary flyby of planet P orbiting Sun S. The planet orbits the Sun at velocity V_{planet}. The initial and final probe velocities relative to the planet are $V_{srp,bef}$ and $V_{srp,aft}$ respectively. (a) The spaceprobe is in an initial retrograde orbit relative to the planet, and the deflection angle is 180°. (b) The spaceprobe is in an initial retrograde orbit relative to the planet, and the deflection angle is 90°.

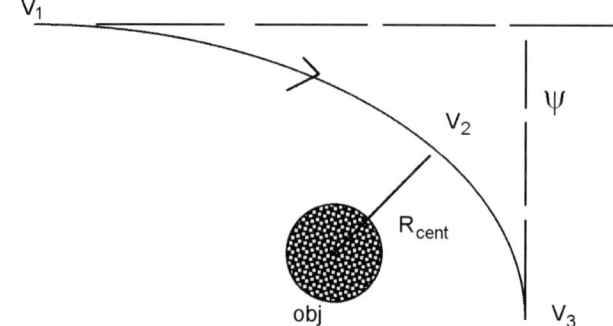

Fig. 4.4. The geometry of a flyby of a celestial object (obj). V_1, pre-periapsis probe velocity relative to obj; V_2, periapsis probe velocity relative to obj; V_3, post-periapsis probe velocity relative to obj; R_{cent}, periapsis separation of probe from centre of obj; ψ, trajectory bend angle (here, 90°). Dotted lines represent asymptotes to pre-periapsis and post-periapsis trajectories.

deflected by 180°, its post-encounter velocity relative to the Sun is $V_{srp,bef} + V_{planet}$. In this extreme case, the spacecraft's velocity relative to the Sun has been increased by $2V_{planet}$. Similar reasoning applied to the 90-degree trajectory deflection of Figure 4.3(b) reveals that in this case the post-encounter velocity is increased by V_{planet}.

Jupiter orbits the Sun at about $13\,\mathrm{km\,s^{-1}}$. A 90-degree unpowered trajectory deflection by Jupiter will therefore increase a spaceprobe's velocity relative to the Sun by $13\,\mathrm{km\,s^{-1}}$. A 180-degree unpowered trajectory deflection by Jupiter will increase the probe's velocity relative to the Sun by $26\,\mathrm{km\,s^{-1}}$.

Further details of unpowered planetary gravity assists are presented in Figure 4.4. The trajectory bend angle is ψ; probe velocities relative to the planet before periapsis (closest approach), at periapsis and after perapsis are respectively V_1, V_2 and V_3; and R_{cent} is the probe's separation from the centre of celestial object 'obj' at periapsis.

One important parameter in estimating gravity-assist kinematics is the parabolic (escape) velocity at the periapsis approach distance to the object. This is approximated as

$$V_{para,obj} \approx 1.4 \left(\frac{GM_{obj}}{R_{cent}} \right)^{1/2} \qquad (4.8)$$

where G is the gravitational constant, and M_{obj} is the mass of the celestial object. The trajectory bend angle, ψ, can be estimated using a modification of equation (4.8) of Flandro's 1966 analysis:

$$\psi = \sin^{-1}\left[\frac{1}{1 + 2(V_1/V_{para.obj})^2} \right] + \sin^{-1}\left[\frac{1}{1 + 2(V_3/V_{para.obj})^2} \right] \qquad (4.9)$$

As an extrasolar probe recedes from the Sun after a planetary gravity assist, its total energy (kinetic + potential energies) in a Sun-centred coordinate system is invariant. The probe's potential energy relative to the Sun increases as its distance from the Sun (R_{sun}) increases. Therefore, its velocity relative to the Sun (V_{prs}) will decrease as the probe travels further from the Sun. By equating total probe energy at

Table 4.2. Periapsis parabolic velocities and angular deflections for an unpowered Jupiter flyby with $V_1 = V_3 = 11\,\text{km}\,\text{s}^{-1}$

Periapsis distance	Periapsis parabolic velocity	Trajectory deflection angle
7.0E4 km	6.0E4 m s^{-1}	139.2°
1.0E5	5.0E4	131.8°
2.0E5	3.6E4	114.2°
3.0E5	2.9E4	102.0°
3.5E5	2.7E4	97.1°
4.0E5	2.5E4	92.7°
4.5E5	2.4E4	88.8°

points 1 and 2, the probe velocity at point 2 can be expressed as a function of its velocity relative to the Sun at point 1 and the probe–Sun separations at points 1 and 2:

$$V_{prs,2} = \sqrt{V_{prs,1}^2 - 2GM_{sun}\left(\frac{1}{R_{sun,1}} - \frac{1}{R_{sun,2}}\right)} \quad (4.10)$$

Exercise 4.3 Applying the standard definitions of kinetic and potential energies, derive equation (4.10).

From Stone and Lane's 1979 paper, Voyager 1 approached Jupiter within about 350,000 km, or 4.8 planetary radii, with $V_1 = V_3 = 11\,\text{km}\,\text{s}^{-1}$. Equations (4.8) and (4.9) have been solved for the case of an $11\,\text{km}\,\text{s}^{-1}$ Jovian flyby, after substituting $G = 6.67 \times 10^{-11}$ Newton metre^2 kg^2, and $M_{\text{Jupiter}} = 1.9 \times 10^{27}$ kg. These results are presented in Table 4.2 for a variety of periapsis distance (measured from the planet's centre). The trajectory bend angle from the Stone and Lane reference was about 98°.

Exercise 4.4 Repeat the calculations used to determine Table 4.2 for the case of an $11\,\text{km}\,\text{s}^{-1}$ unpowered approach to Saturn, which has a mass about 30% that of Jupiter's. Then investigate the effect of varying the value of probe approach velocity relative to the planet, for both Jupiter and Saturn.

Assume that a probe's trajectory bend angle is 90° after an $11\,\text{km}\,\text{s}^{-1}$ Jupiter flyby. From the above discussion, the probe's velocity relative to the Sun will be about $24\,\text{km}\,\text{s}^{-1}$, at Jupiter's orbit roughly 5 AU from the Sun. Equation (4.10) and the solar mass of about 2×10^{30} kg can be used to estimate the Solar System exit velocity (at infinite solar distance) for the probe as $14.8\,\text{km}\,\text{s}^{-1}$ or about 3.1 AU per year.

Exercise 4.5 Apply equation (4.10) for the above probe to plot a curve of probe speed relative to the Sun against probe distance from the Sun between probe–Sun separations of 5 and 50 AU.

4.4 POWERED SOLAR GRAVITY ASSISTS

Even if we customise our probe's trajectory so that it passes close to all of the Jovian planets, as did Voyager 2, it will be difficult to greatly exceed Voyager 2's Solar System exit velocity of 3.5 AU per year using unpowered planetary flybys. At 3.5 AU per year, the 4.3 light year journey to our Sun's nearest stellar neighbour would take approximately 80,000 years. Unpowered planetary flybys, while a good technique for the exploration of the outer Solar System and the nearby interstellar medium, is totally inadequate for the accomplishment of true interstellar travel.

Another approach is to direct the spacecraft deep into the gravity well of a celestial object and then perform an impulsive powered manoeuvre along the line of flight at or near periapsis. One of the first researchers to consider such a manoeuvre was Kraft Ehricke, in 1972.

Ehricke's proposal used multiple swingbys of Jupiter and Saturn to direct a spacecraft onto a parabolic or slightly hyperbolic solar orbit with a perihelion 0.01–0.03 AU from the Sun's centre. For Solar System exit velocities as high as 0.003 c, 50% or less of the propulsive energy must be supplied by the ship's onboard motors.

Because of a misleading statement in Ehricke's treatment, we follow instead the 1986 analysis of Matloff and Parks. Consider the spaceprobe flyby configuration represented by Figure 4.4. Although the powered-flyby technique will work for any celestial object, we here consider that the celestial object targeted is our Sun, the most massive object in the Solar System. The pre-periapsis and post-periapsis points (corresponding to V_1 and V_3 respectively in Figure 4.4) are here considered to be very much further from the Sun than the perihelion point V_2. Assuming that the probe–Sun pre-perihelion and post-perihelion distances are essentially infinite, and defining the perihelion probe separation from the Sun's centre as R_{peri}, it is immediately possible to define the ratio of total probe energy ($TE =$ kinetic + potential energies) to probe mass at the points corresponding to V_1, V_2, and V_3:

$$\left(\frac{TE}{M_{\text{probe}}}\right)_{\text{pre-perihelion}} = \frac{1}{2}V_1^2$$

$$\left(\frac{TE}{M_{\text{probe}}}\right)_{\text{perihelion}} = \frac{1}{2}V_2^2 - \frac{GM_{\text{sun}}}{R_{\text{peri}}} \quad (4.11)$$

$$\left(\frac{TE}{M_{\text{probe}}}\right)_{\text{post-perihelion}} = \frac{1}{2}V_3^2$$

If we next equate total energy at the pre-perihelion point to the total energy just before the motors are fired at perihelion, and refer to the definition of parabolic velocity in equation (4.8), the pre-burn perihelion velocity of the probe relative to the Sun can be expressed as

$$V_2 = \sqrt{V_1^2 + V_{\text{para-peri}}^2} \quad (4.12)$$

where $V_{\text{para-peri}}$ is the Sun's parabolic or escape velocity at the probe's perihelion distance.

Now, if there is an impulsive velocity change at perihelion, ΔV_{peri}, the kinetic energy per probe mass just after this velocity increase can be written as $\frac{1}{2}(V_2 + \Delta V_{peri})^2$. If we next equate the total post-impulse energy/mass at perihelion to the post-perihelion total energy/mass at the post-perihelion point, we can apply the definition of parabolic velocity to obtain an expression for the probe's velocity relative to the Sun at the post-perihelion point:

$$V_3 = \left[\left(\sqrt{V_1^2 + V_{para-peri}^2} + \Delta V_{peri} \right)^2 - V_{para-peri}^2 \right]^{1/2} \quad (4.13)$$

The trajectory deflection angle for the case of a powered perihelion manoeuvre can be estimated using equation (4.9).

If we next assume that the parabolic velocity at perihelion is very much greater than both the impulsive velocity increment at perihelion and the pre-perihelion velocity, we obtain the following approximation (after some manipulation) for post-perihelion probe velocity:

$$V_3 \approx \sqrt{V_1^2 + \Delta V_{peri}^2 + 2\Delta V_{peri} V_{para-peri}} \quad (4.14)$$

The difference between post-perihelion velocity and pre-perihelion velocity can now be approximated:

$$\Delta V_{post-peri} \approx V_3 - V_1 = \sqrt{V_1^2 + 2\Delta V_{peri} V_{para-peri}} - V_1 \quad (4.15)$$

In the special case of a parabolic pre-perihelion trajectory, $V_1 = 0$ and $V_3 = \Delta V_{post-peri} \approx (2\Delta V_{peri} V_{para-peri})^{1/2}$.

To investigate the utility of a powered solar flyby, consider the case of a space probe that flies by the Sun at a distance of 0.01 AU from the Sun's centre (about two solar radii). At perihelion, the probe's velocity is increased by $2\,km\,s^{-1}$. Application of equation (4.8) reveals that the Sun's escape velocity at 0.01 AU from the Sun's centre is about $420\,km\,s^{-1}$. If the probe accelerates by $2\,km\,s^{-1}$ at perihelion, equation (4.15) reveals that the probe leaves the Solar System with a velocity of about $41\,km\,s^{-1}$. This converts to 8.7 AU per year – almost three times the Solar System exit velocities of the Voyager probes.

Exercise 4.6 Verify all steps in the derivation of equation (4.15). For the example just considered, plot a curve of Solar System exit velocity against perihelion velocity increase, for perihelion velocity increases between 0 and $10\,km\,s^{-1}$.

The potential of planetary and solar gravity assist manoeuvres has not been exhausted by the examples presented here. Sophisticated computer analysis has led to a whole series of planetary and extraplanetary applications of these techniques.

Claudio Maccone has recently demonstrated that two optimised Jupiter flybys and one intermediate Sun flyby can eject a spacecraft from the Solar System at about $51\,km\,s^{-1}$. Including the time required for manoeuvres within our planetary system, an extrasolar probe using this technique can reach a point near the ecliptic and about 550 AU from the Sun within a human lifetime.

A totally new approach to celestial dynamics using much more advanced computational techniques has been applied by Edward Belbruno to develop a family of low-energy trajectories for use in the Earth–Moon system. One of these 'fuzzy boundary' trajectories was applied to move the Japanese probe Hiten from Earth-orbit to the lunar vicinity in 1991, with a minimum propellant requirement.

4.5 THE SOLAR-PHOTON SAIL

Although this approach to spaceflight has a venerable theoretical history, the first test of a solar-photon sail in space did not occur until 1993, when Russian engineers unfurled Znamia 2, an experimental thin-film sunlight reflector, from the space station Mir. Although the theory was pioneered by Russian scientists in the first half of the twentieth century, the most significant early theoretical paper dealing with solar sailing in the solar system was by Tsu, in 1959.

In the late 1970s, solar-sail design moved into high gear as a NASA/JPL research team directed by Louis Friedman considered robot sails for the later-cancelled 1986 US mission to Halley's comet. (This work and the early history of solar-sail research is reviewed in cited references by Friedman, Polyakhova and Wright.)

The basic principle of solar sailing can be understood with the aid of Figure 4.5. Sunlight impinges on the sail – a highly reflective, very thin film. In most designs, the sail is connected to the payload by super-strong cables constructed from industrial diamond or silicon carbide (SiC) filaments. Additional structural elements might include a thin-film mesh attached to the front (anti-sunward) sail face which serves as a ripstop in case of micrometeroid impacts. Note the three basic sail designs. In a parachute-type sail, the solar radiation pressure supports the sail against the retarding masses of payload and cables. The payload and cables trail the parachute sail. A hollow-body or 'pillow-type' sail is inflated with gas and dispenses with the cable (Bernasconi/Reibaldi and Strobl). The payload of a hollow-body sail is placed in front of the sail. A parabolic-type sail is similar to the parachute-type except that its shape is adjusted to focus reflected sunlight on a small, steerable thin-film reflector closer to the Sun than is the payload. This reflector can vector reflected sunlight in a variety of directions, allowing the parabolic-type sail to operate as a highly manoeuverable 'solar-photon thruster.'

Basic sail theory

Propulsion is accomplished by the transfer of momentum from reflected photons to the sail. This solar radiation pressure on an opaque (non-transmissive) sail is expressed as

$$RP_{\text{sail}} = \frac{(1 + REF_{\text{sail}})}{c} S_c \text{ N m}^{-2} \qquad (4.16)$$

where REF_{sail} is the sail reflectivity, c is the speed of light, and S_c refers to the solar energy per second incident on the sail ($S_c = 1400 \, R_{au}^2 \text{ W m}^{-2}$, where R_{au} is the distance to the Sun in Astronomical Units).

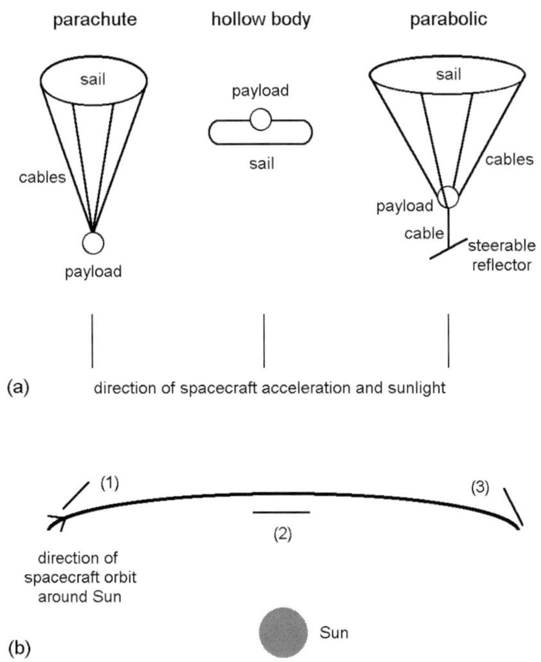

The solar-photon sail

Fig. 4.5. Variations of the solar-photon sail: (a) three types of sail; (b) three sail orientations. In (b), at position (1) the component of solar radiation pressure is tangential to the orbital path in the opposite direction from the orbit, and the sail moves closer to the Sun; at position (2) the sail is normal to the direction of the sunlight; at position (3) the component of solar radiation pressure is tangential to the orbital path in the same direction as the orbit, and the sail moves further from the Sun.

The ship mass M_s is the sum of sail, cable, structural and payload masses, and the area of a disc-shaped sail is written as πR_{sail}^2, where R_{sail} is the sail's radius. Applying the definitions of force and acceleration in equation (4.16), we find that the sail's acceleration due to solar radiation pressure is

$$ACC_{rp,sail} = \frac{(1 + REF_{sail})}{cM_s} S_c \pi R_{sail}^2 \text{ m/sec}^2 \qquad (4.17)$$

We next define the sail lightness factor, η_{sail}, which is the ratio of solar radiation pressure acceleration on the sail to solar gravitational acceleration on the sail. If ship mass is redefined as $M_s = \sigma_{eff} \pi R_{sail}^2$, where σ_{eff} is the effective areal mass thickness of the ship (ship mass in kg per sail area in square metres), the sail lightness factor can be written as

$$\eta_{sail} = \frac{(1 + REF_{sail})}{c\sigma_{eff} GM_{sun}} S_c R_{sun}^2 \qquad (4.18)$$

where G is the gravitational constant, M_{sun} is the Sun's mass and R_{sun} is the distance between the sail and the Sun's centre.

By converting Astronomical Units to metres, it is easy to show that the term S_c in equation (4.18) can be expressed as $3.15 \times 10^{25} \, R_{\text{sun}}^{-2}$. Substituting this factor into equation (4.18), and the MKS values for the speed of light, the mass of the Sun and the gravitational constant, we obtain an expression for lightness factor that is independent of the distance to the Sun:

$$\eta_{\text{sail}} = 7.87 \times 10^{-4} \left(\frac{1 + REF_{\text{sail}}}{\sigma_{\text{eff}}} \right) \quad (4.19)$$

Although useful as a close approximation, this result must be refined for space-manufactured sails so thin that they are partially transmissive. In such cases reflectivity and other optical constants can be calculated by the method outlined by Matloff in a 1984 *Journal of the British Interplanetary Society* paper and a 1997 IAA paper. Forward's 1990 paper on 'Grey Solar Sails' also deals with the accurate calculation of sail optical properties.

Sail thermal effects

As indicated by equation (4.17), sail radiation-pressure acceleration increases as the spacecraft approaches the Sun. It is easy to derive the thermal limitations for a sail operating close to the Sun.

Consider an 'inward-bent' sail such as the one shown in Figure 4.6. This sail gradually unfurls as it recedes from the Sun, so as to reduce acceleration load on the payload. Although fully unfurled at perihelion, only a fraction f_u of its area at perihelion is directed normal to the Sun (Figure 4.5(b) (3)). The sail area normal to the Sun at perihelion is $f_u \pi R_{\text{sail}}^2$. The solar energy per second incident on the sail at perihelion is S_c multiplied by this effective perihelion sail area. The solar energy per second absorbed by the (opaque) sail at perihelion can be directly expressed as

$$P_{abs} = \frac{1400}{R_{\text{peri},au}^2} f_u \pi R_{\text{sail}}^2 (1 - REF_{\text{sail}}) \text{ W} \quad (4.20)$$

where $R_{\text{peri},au}$ is the distance in Astronomical Units between the sail and the Sun's centre at perihelion.

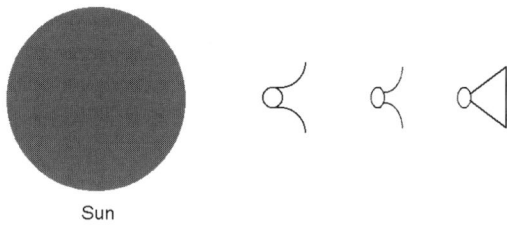

Fig. 4.6. Gradual unfurlment of an inward-bent solar sail as it recedes from the Sun.

From elementary radiation theory, the sail radiant emittance (W_{em}) in W m^{-2} at perihelion is P_{abs} divided by the factor ($2\pi R_{sail}^2$). The factor '2' arises because the sail can radiate from both sides. Applying the Stefan–Boltzmann law for grey bodies ($W_{em} = \varepsilon\sigma T(k)^4$) where ε is the sail's emissivity (1- REF_{sail} for a non-transmissive sail), σ is the Stefan-Boltzmann constant (5.67 × 10^{-8} in MKS units), and $T(k)$ is the absolute or Kelvin temperature of the sail material. The sail temperature at perihelion is easily determined to be

$$T(k) = 333 \left[\frac{f_u}{\varepsilon R_{peri,au}^2} (1 - REF_{sail}) \right]^{1/4} \text{ K} \quad (4.21)$$

This result is close to the expression calculated by Koblik *et al.*

Cable stress considerations

Applying membrane theory, but actually repeating D'Alembert's eighteenth century analysis of the parachute, Matloff and Mallove estimated parachute-sail-type cable mass in their 1981 *Journal of the British Interplanetary Society* paper. The required cable mass is written:

$$M_{cable} = \frac{1.4\rho_{cable}R_{sail}M_{payload}}{\frac{(TS)_{cable}}{Acc_{sail,max}} - 1.4\rho_{cable}R_{sail}} \text{ kg} \quad (4.22)$$

where ρ_{cable} is the cable density, $(TS)_{cable}$ is the cable tensile strength, and $Acc_{sail,max}$ is the maximum spacecraft acceleration.

This approximation has been refined by Ewing in 1992 and by Cassenti *et al.* in 1996. As shown by Matloff in 1996, equation (4.22) overestimates cable mass in the case of very-thin, highly reflective cables that are affected by solar radiation pressure during a close perihelion pass.

Simplified interstellar sail kinematics

Solar System operation of solar-photon sails is described by Tsu. To direct a solar sail from the vicinity of Earth to that of Mars, the sail is oriented with a positive aspect angle (as shown in Figure 4.5(b)), so that a component of the radiation-pressure force on the sail is parallel to and in the same direction as the spacecraft's velocity vector. To spiral back to the vicinity of the Earth from the vicinity of Mars, the sail is oriented so that a component of the radiation pressure force opposes the spacecraft's velocity vector.

One approach to interstellar solar sailing – developed theoretically by Matloff and Mallove in 1981 – is shown schematically in Figure 4.7(a). The partially unfurled sail is placed behind a much more massive asteroidal occulter. Occulter and starship initially follow a parabolic or slightly hyperbolic solar orbit, with a perihelion as close as 0.01 AU from the Sun's centre. During the pre-perihelion phase, the occulter protects the sail from premature ejection from the Solar System caused by solar radiation pressure.

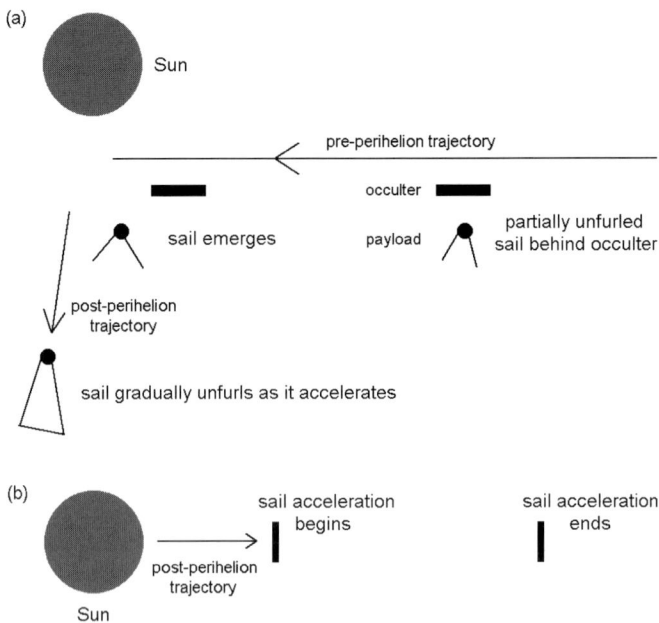

Fig. 4.7. Two approaches to interstellar solar sailing: (a) gradual sail unfurlment after perihelion; (b) sail fully unfurled at perihelion.

At perihelion, the sail (which is oriented normal to the Sun) emerges from behind the occulter, which continues along the original trajectory. The sail gradually unfurls as it accelerates from the Solar System. In some cases, ballast is also released in order to maintain constant acceleration for as long as possible. In 1983, Matloff and Mallove published in *Journal of the British Interplanetary Society* a computer program to optimise mission design parameters as a function of sail material, cable tensile strength and density, payload mass and other factors. Many configurations proved capable of projecting a small human-occupied starship (in the mult-million-kilogramme range) towards α Centauri on voyages of a millennium or less.

A follow-on 1983 *Journal of the British Interplanetary Society* paper by Matloff investigated various methods of pre-perihelion acceleration to reduce travel time. Forward's solar-photon-thruster concept (the parabolic sail shown in Figure 4.5) might conceiveably eliminate the requirement for the occulter.

In 1996 and 1997, papers by Brice Cassenti and Giovanni Vulpetti presented computer analysis of pre-perihelion sail trajectory optimisation. Although normal sail orientation to the Sun is an easier case to analyse, optimised sail angles relative to the Sun result in somewhat better performance.

At the STAIF 2000 conference, Vulpetti presented a class of trajectories using non-normal sail orientation with near-term application to interstellar precursor probes and ultimate application to true solar-sail starships. Assuming a 345-kg sailcraft with a payload of 100 kg and an aluminium–chromium sail with an area of $0.287\,\mathrm{km}^2$, a 0.151-AU perihelion distance is sufficient to obtain a Solar System exit velocity of 25 AU per year. About 114 days of pre-perihelion manoeuvre are required, and there are no other propulsion systems or lunar/planetary launch window constraints. If necessary, the mission could be aborted in the Solar System and the craft returned to Earth prior to the perihelion passage.

Analytical work to date has assumed that the amount of solar energy per second incident upon a sail decreases with the inverse square of the distance to the Sun's centre. Since the Sun is an extended source, as opposed to a point light source, McInnes/Brown and Shvartsburg have argued that the inverse-square assumption is not exact for very close perihelion distances.

The following is an approximate analytical approach to interstellar solar sailing, based upon a 1991 *Journal of the British Interplanetary Society* paper by Matloff. Although much mathematical complexity is avoided, it must be re-emphasised that the approach is not exact. As shown in Figure 4.7(b), a fully unfurled sail is assumed at perihelion, the sail is opaque and oriented normal to the Sun, and the pre-perihelion trajectory is a parabolic solar orbit.

If no ballast is released, the ratio of total energy to mass is identical at the sail unfurlment point (here called 'R_{init}') and at the point where sail acceleration is terminated (here called 'R_{fin}'). The change in spacecraft kinetic energy per mass can be defined as the work per spacecraft mass done on the sail by the Sun's radiation pressure force as the distance from the Sun varies between R_{init} and R_{fin}:

$$\Delta KE/M_s = \int_{R_{init}}^{R_{fin}} ACC_{rp,sail} dR_{sun} \quad (4.23)$$

(Note that we are using a sun-centred coordinate system, directed outward from the Sun.) From the definition of sail lightness factor η_{sail},

$$ACC_{rp,sail} = \eta_{sail} G \frac{M_{sun}}{R_{sun}^2} \quad (4.24)$$

Substituting equation (4.24) into equation (4.23) and integrating, we obtain

$$\Delta KE/M_s = \eta_{sail} G M_{sun} \left(\frac{1}{R_{init}} - \frac{1}{R_{fin}} \right) \quad (4.25)$$

If we use equation (4.25), the definitions of potential energy/spacecraft mass and solar parabolic (escape) velocity in an equation equating total energies at R_{init} and R_{fin}, we obtain the following relationship for the final spacecraft velocity relative to the Sun:

$$V_{fin} = [V_{init}^2 + (\eta_{sail} - 1)(V_{para,init}^2 - V_{para,fin}^2)]^{1/2} \quad (4.26)$$

where V_{init} is the sail velocity relative to the Sun at sail unfurlment, and the subscript 'para' refers to solar parabolic velocities.

If the ship is in an initially parabolic solar orbit, $V_{init} = V_{para}$, init $= V_{para-peri}$, and acceleration ends far from the Sun:

$$V_{fin} \approx \eta_{sail}^{1/2} V_{para-peri} \quad (4.27)$$

Exercise 4.7 Verify all steps in the derivation of equation (4.27).

Within the vicinity of the inner planets of the Solar System, even very thin solar sails will be low-thrust devices. Since the Sun's gravitational field accelerates the Earth by about 6×10^{-4} g (0.006 m s^{-2}), a sailing ship with a lightness factor of 10 will experience a solar-radiation-pressure acceleration of about 0.006 g at 1 AU from the Sun if the sail is fully unfurled and oriented normal to the Sun. But during a 0.01 AU perihelion pass, the same sail will experience a staggering 60 g! Partial sail-unfurlment and the use of ballast will be necessary to moderate perihelion acceleration, at least if we ever launch crewed solar-sail starships like the fictional starship in the novel by Apollo 11 astronaut Buzz Aldrin and John Barnes (*Encounter with Tiber*, 1996).

The high perihelion accelerations is also why crewed ships must use superstrong and low-density cable material. The Matloff/Mallove (1981) proposal uses diamond cables with a density of $3{,}520 \text{ kg m}^{-3}$ and a tensile strength of $5.3 \times 10^{10} \text{ N m}^{-2}$.

Performance of robotic extrasolar or interstellar sails can be evaluated more easily than that of crewed ships using the above analysis. Consider, for example, a 30-nm aluminium sail with a reflectivity of 0.9. (This is about the thinnest possible fully opaque aluminium sail). The density of aluminium is $2{,}700 \text{ kg m}^{-3}$, and its melting point is 933 K.

If our 30-nm aluminium sail is fully unfurled at perihelion, is oriented normal to the Sun and has a maximum operating temperature of 900 K, the emissivity of the fully opaque sail is 0.1. Substituting in equation (4.21), the perihelion distance is estimated to be 0.13 AU. (This example is for illustrative purposes only. It is unlikely that pure aluminium sails will function at temperatures higher than 600 K.)

The areal mass thickness of our 30-nm aluminium sail is calculated to be $8.1 \times 10^{-5} \text{ kg m}^{-3}$. Substituting into equation (4.19), this sail's lightness factor is 18.5.

At 1 AU from the Sun, the solar parabolic velocity is about 42 km s^{-1}. From equation (4.8), solar parabolic velocity scales with the inverse square root of perihelion distance. At 0.13 AU from the Sun's centre, the parabolic velocity is 116 km s^{-1}. Substituting $V_{\text{para-peri}} = 116 \text{ km s}^{-1}$ and $\eta_{\text{sail}} = 18.5$ into equation (4.27), we obtain a Solar System exit velocity of about 500 km s^{-1}.

This velocity converts to 105 AU per year – about 30 times the velocity of Voyagers 1 and 2. Our hypothetical interstellar probe will reach α Centauri in 'only' about 2,500 years.

Exercise 4.8 The above calculation did not consider either payload mass or more advanced sail materials. Assume that the same sail is 'rigged' as a hollow-body sail with a payload that raises the areal mass thickness to $10^{-4} \text{ kg m}^{-3}$, and repeat the calculation of Solar System exit velocity. Then try it again for the case of a thinner sail with identical thermal, mechanical and optical properties such that the areal mass thickness is $5 \times 10^{-4} \text{ kg m}^{-3}$.

Matloff in a 1997 IAA paper, and Landis in a 1997 *Journal of the British Interplanetary Society* paper, describe studies of sail material indicating that partially-transparent 20-nm-thick beryllium is superior to other metallic monolayers for interstellar solar-photon sailing application. Matloff and Mallove, in their 1981 and 1983 *Journal of the British Interplanetary Society* papers, also consider bilayers, with a highly reflective metallic layer facing towards the Sun and an emissive layer comprising the anti-sunward face of the sail.

After acceleration, the sail (and cables) could be wound around the payload to serve as cosmic ray shielding during the long interstellar transfer. The sail could be redeployed as the destination star is approached, and used in reverse as a deceleration mechanism.

A rigorous approach to starship deceleration by solar sail is derived in the 1985 *Journal of the British Interplanetary Society* paper by Matloff and Ubell, and is reviewed in *The Starflight Handbook*. A simpler approximation – derived in a manner analogous to equation (4.27) – is included in the 1991 AIAA paper by Matloff *et al.*

Assume with Matloff *et al.* (1991) that a starship cruises towards the destination star at a velocity $V_{cr,\text{init}}$ (relative to the destination star) before the sail is unfurled for deceleration. If the sail is unfurled and used to decelerate the starship to $V_{cr,\text{fin}}$ at a distance R_{fin} from the centre of the destination star, it is easy to show that

$$V_{cr,\text{init}} \approx [V_{cr,\text{fin}}^2 + (\eta_{\text{sail}} - 1)V_{\text{para}-R_{\text{fin}}}^2]^{1/2} \qquad (4.28)$$

where $V_{\text{para},R_{\text{fin}}}$ is the parabolic or escape velocity at a distance R_{fin} from the centre of the destination star.

Earth-launched current-technology sails of the 'Znamia' class are typically several microns thick. Friedman speculates that Earth-launched sails of 1 μm or so in thickness are not impossible. In 1996, Santoli and Scaglione speculated that the thickness of an Earth-launched sail could be reduced to a fraction of a micron in orbit if the sail consists of an aluminium/plastic bilayer in which the plastic substrate is designed to evaporate when exposed to solar ultraviolet radiation. But for the production of sails as thin as 20 or 30 nm, space manufacturing of sail film is a requirement. But as described in the previous chapter, a recent advance in materials technology may lead to interconnected-mesh Earth-launched sails about 50% as effective as the best conceivable space-manufactured hyperthin metallic sheet sails. New approches to sail stress analysis will also result in innovative sail designs and unfurlment strategies. (See, for example, the cited works by Genta and Brusca.)

This propulsion system has great promise, at least for early robotic forays into nearby interstellar space. Whether there will ultimately be human-inhabited interstellar arks propelled from the Sun on millennia-duration journies by solar-photon sails is a question that will be answered in the future.

As discussed by Matloff and Mallove in 1983, the utility of the interstellar solar sail will increase in the very far future, just when humanity (or its descendants) most requires such a technology. As the Sun leaves the main sequence in about 5 billion years, it will expand towards the giant phase on the Hertzsprung–Russell diagram. Since the solar constant will then greatly increase, migrating interstellar starships propelled by solar sails will be considerably faster than those launched from the present-day Sun. Voyages that today might require 1,000 years or more might have durations of only a few centuries, in that far-distant era.

4.6 BIBLIOGRAPHY

'Deep Space 1 Spacecraft', noted in *Spaceflight*, **40**, 352 (1998).

Adams, C. S., *Space Flight*, McGraw-Hill, New York (1958).

Belbruno, E., 'Through the Fuzzy Boundary: A New Route to the Moon', *The Planetary Report*, **12**, No. 3, 8–10 (1992).

Belbruno, E. and Miller, J. K., 'Sun-Perturbed Earth-to-Moon Transfers with Ballistic Capture', *Journal of Guidance, Control and Dynamics*, **16**, 770–776 (1993).

Bernasconi, M. C. and Reibaldi, G. C., 'Inflatable Space-Rigidized Structures: Overview of Applications and Technology Impact', *Acta Astronautica*, **14**, 455–465 (1986).

Brewer, G. R., *Ion Propulsion*, Gordon and Breach, Philadelphia, PA (1970).

Casani, E. K., Stocky, J. F. and Rayman, M. D., 'Solar Electric Propulsion', *Missions to the Outer Solar System and Beyond, 1st IAA Symposium on Realistic Near-Term Scientific Space Missions*, ed. G. Genta, Levrotto & Bella, Turin, Italy (1996), pp. 143–158.

Cassenti, B. N., Matloff, G. L. and Strobl, J., 'The Structural Response and Stability of Interstellar Solar Sails,' *Journal of the British Interplanetary Society*, **49**, 345–350 (1996).

Cassenti, B. N., 'Optimization of Interstellar Solar Sail Velocities', *Journal of the British Interplanetary Society*, **50**, 475–478 (1997).

Ehricke, K. E., 'Saturn-Jupiter Rebound: A Method of High Speed Spacecraft Ejection from the Solar system', *Journal of the British Interplanetary Society*, **25**, 561–572 (1972).

Ewing, A., 'Solar Sail Spacecraft Design using Dimensional Analysis', IAA-92-0239.

Flandro, G. A., 'Fast Reconaissance Missions to the Outer Solar System Utilizing Energy Derived from the Gravitational Field of Jupiter', *Astronautica Acta*, **12**, 329–337 (1966).

Friedman, L., *Starsailing*, Wiley, New York (1988).

Forward, R. L., 'Grey Solar Sails', *The Journal of the Astronautical Sciences*, **38**, 161–185 (1990).

Forward, R. L., 'Solar Photon Thruster', *Journal of Spacecraft*, **27**, 411–416 (1990).

Forward, R. L. and Davis, J., *Mirror Matter*, Wiley, New York (1988).

Genta, G. and Brusca, E., 'The Parachute Sail with Hydrostatic Beam: A New Concept for Solar Sailing', and 'The Aurora Project: A New Sail Layout', in *Missions to the Outer Solar System and Beyond, 2nd IAA Symposium on Realistic Near-Term Scientific Space Missions*, ed. G. Genta, Levrotto & Bella, Turin, Italy (1988), pp. 61–68 and 69–74. Also published in *Acta Astronautica*, **44**, 133–140 and 141–146 (1999).

Joels, K. M., Kennedy, G. P. and Larkin, D., *The Space Shuttle Operator's Manual*, Ballantine, New York (1982).

Koblik, V. V., Polyakhova, E. N. Sokolov, L. L. and Shmyrov, A. S., 'Controlled Solar Sailing Transfer Flights into Near-Sun Orbits under Restrictions on Sail Temperature', *Cosmic Research*, **34**, 572–578 (1996).

Landis, G. A., 'Photovoltaic Receivers for Laser Beamed Power in Space', NASA Contract Report 189075, Sverdrup Technology Inc., Lewis Research Center Group, Brook Park, OH, USA.

Landis, G. A., 'Small Laser-Pushed Lightsail Interstellar Probe', *Journal of the British Interplanetary Society*, **50**, 149–154 (1997).

Maccone, C., 'Solar Foci Missions', IAA-L-0604, in *Proceedings of the Second International Conference on Low-Cost Planetary Missions*, Laurel, MD, April 1996.

Mallove, E. F. and Matloff, G. L., *The Starflight Handbook*, Wiley, New York (1989).

Matloff, G. L. and Mallove, E. F., 'Solar Sail Starships – the Clipper Ships of the Galaxy', *Journal of the British Interplanetary Society*, **34**, 371–380 (1981).

Matloff, G. L. and Mallove, E. F., 'The Interstellar Solar Sail: Optimization and Further Analysis', *Journal of the British Interplanetary Society*, **36**, 201–209 (1983).

Matloff, G. L., 'Beyond the Thousand Year Ark: Further study of Non-Nuclear Interstellar Flight', *Journal of the British Interplanetary Society*, **36**, 483–489 (1983).

Matloff, G. L., 'Interstellar Solar Sailing: Consideration of Real and Projected Sail Material', *Journal of the British Interplanetary Society*, **37**, 135–141 (1984).

Matloff, G. L. and Ubell, C. B., 'Worldships: Prospects for Non-Nuclear Propulsion and Power Sources', *Journal of the British Interplanetary Society*, **38**, 253–261 (1985).

Matloff, G. L. and Parks, K., 'Interstellar Gravity Assist Propulsion: A Correction and New Application', *Journal of the British Interplanetary Society*, **41**, 519–526 (1986).

Matloff, G. L., 'Early Interstellar Precursor Solar Sail Probes', *Journal of the British Interplanetary Society*, **44**, 367–370 (1991).

Matloff, G. L., Walker, E. H. and Parks, K., 'Interstellar Solar Sailing: Application of Electrodynamic Turning', AIAA-91-2538.

Matloff, G. L., 'The Impact of Nanotechnology upon Interstellar Solar Sailing and SETI', *Journal of the British Interplanetary Society*, **49**, 307–312 (1996).

Matloff, G. L., 'Interstellar Solar Sails: Projected Performance of Partially Transmissive Sail Films', IAA-97-IAA.4.1.04.

Mauldin, J. H., *Prospects for Interstellar Travel*, Univelt, San Diego, CA (1992).

McInnes, C. R. and Brown, J. C., 'Solar Sail Dynamics with an Extended Source of Radiation Pressure', *Acta Astronautica*, **22**, 155–160 (1990).

McInnes, C. R., *Solar Sailing*, Praxis, Chichester, UK (1999), pp. 43–46.

Nock, K., 'TAU – A Mission to a Thousand Astronomical Units', AIAA-87-1049.

Polyakhova, E. N., *Spaceflight Using a Solar Sail: The Problems and Prospects* (in Russian), Vol 9., Mekhanika Kismicheskogo Poleta Series, Isdatelstvo Nauka (1986). (English translation JPRS-USP-88-003-L available to US Government employees and contractors.)

Potter, S. D., 'Applications of Thin-Film Technology in Space Power Systems', in Proceedings of High Frontier Conference XII, Space Studies Institute, Princeton, NJ (May 4–7, 1995).

Ravenni, L., 'Flyby: Una Spinta Per Esplorare L'Universo', thesis (in Italian), Università degli Studi di Siena, Siena, Italy (1997).

Santoli, S. and Scaglione, S., 'Project Aurora: A Preliminary Study of a Light, All-Metal Solar Sail', in *Missions to the Outer Solar System and Beyond, 1st IAA Symposium on Realistic Near-Term Scientific Space Missions*, ed. G. Genta, Levrotto & Bella, Turin, Italy (1996), pp. 37–48.

Shvartsburg, A., 'Solar Sail dynamics using Extended Light Source', in *Proceedings of 1993 AINA Conference – Advances in Nonlinear Astrodynamics*, ed. E. Belbruno, Geometry Center, University of Minnesota, Minneapolis, MN (November 8–9, 1993).

Stone, E. C. and Lane, A. L., 'Voyager 1 Encounter with the Jovian System', *Science*, **204**, 945–948 (1979).

Strobl, J., 'The Hollow Body Solar Sail', *Journal of the British Interplanetary Society*, **42**, 515–520 (1989).

Tsu, T. S., 'Interplanetary Travel by Solar Sail', *ARS Journal*, **29**, 422–427 (1959).

Vulpetti, G., '3D High-Speed Escape Heliocentric Trajectories for All-Metal-Sail, Low-Mass Sailcraft', *Acta Astronautica*, **39**, 161–170 (1996).

Vulpetti, G., 'Sailcraft at High Speed by Orbital Angular Momentum Reversal', *Acta Astronautica*, **40**, 733–758 (1997).

Vulpetti, G., 'Sailcraft-Based Mission to the Solar Gravitational Lens', presented at STAIF 2000 Conference, University of New Mexico, Albuquerque, NM, January 30–February 3, 2000.

Wright, J. L., *Space Sailing*, Gordon and Breach, Philadelphia, PA (1992).

5

The incredible shrinking spaceprobe

And so it was indeed: she was now only ten inches high, and her face brightened up at the thought that she was now the right size for going through the little door into that lovely garden.

Lewis Carroll, *Alice in Wonderland* (1865)

If we wish to emulate Alice and venture far into the garden of interstellar vastness, it may be necessary to shrink our spaceprobes – and possibly even the passengers of later interstellar colonisation ships! We can see the necessity for such drastic measures by investigating kinetic energy requirements for a 'modest' interstellar probe.

Consider, with Curt Mileikowsky, the case of a 1,000-kg interstellar probe, launched on an undecelerated flythrough mission to another solar system at a velocity of 0.3 c (about 10^5 km s^{-1}) relative to the Sun. Such a craft would require 12 years to reach α Centauri, not including acceleration time. Because of the limitation of the speed of light, data from this nearest extrasolar neighbour would not reach Earth-bound radio telescopes for another four years, so the total mission duration would be about two decades.

The total kinetic energy of this spacecraft, relative to the Sun, is 5×10^{18} J. If all the electrical power produced by the US power grid (about 10^{13} W) were applied to this mission at 50% efficiency (a very optimistic efficiency scenario), all US power must be beamed to the spacecraft for about 12 days.

An electrical power of 10^{13} W applied for 12 days corresponds to about 3×10^{12} kW hours (KWH). From the author's most recent electricity bill, the cost of electrical energy in New York City is about \$0.13/KWH. The electrical energy cost to accelerate our interstellar spaceprobe will be approximately \$$4 \times 10^{11}$ – a hefty fraction of the US Gross National Product!

Exercise 5.1 Estimate the energy cost of a 1,000-kg undecelerated starprobe cruising at 0.3 c, using Milekowsky's more optimistic electric power cost estimate of $0.03/KWH, for the cases of 10%, 1% and 0.1% efficiency of converting electrical energy into spacecraft kinetic energy.

As we shall see in following chapters, feasible methods of converting electrical energy into spacecraft kinetic energy – laser radiation pressure and antimatter/matter annihiliation – are very inefficient in terms of energy consumption. Electrical costs for a single 1,000-kg starprobe might bankrupt our entire planet!

We have two choices to reduce these costs. The first is to reduce probe velocity and therefore energy requirements. There are obvious limits to this approach. Although a 1,000-year interstellar crossing might not inconvenience the inhabitants of a large interstellar ark or world ship, such a long travel time would upset the science team launching the probe, since none of them would live to see the conclusion of the mission.

The second option is to shrink the probe. This applies the new techniques of nanotechnology, and may ultimately reduce the mass of our 1,000-kg test probe to 1 kg or less.

5.1 THE SMALL AND THE VERY SMALL

As presented in Figure 5.1, spaceprobes have been shrinking in size and cost for several decades. Consider first the 'macro' or 1970s vintage probes. The Viking probes to Mars are good examples of macro spacecraft. As described by Bill Yenne, the mass of each of the two Viking orbiters was about 2,300 kg; each of the two Viking landers that performed life-detection experiments on the martian surface had a mass of about 1,200 kg; and the total Viking 1/2 mission cost was in the neighbourhood of $1 billion US.

Since Viking, robotic spacecraft have shrunk in both size and cost, but not in capabilities. The total mass of mid-1990s 'mini' interplanetary probes, such as the Mars Pathfinder lander, is in the vicinity of 100 kg, and the mission costs have been reduced by a factor of about 10. As discussed by Darren Burnham and Andy Salmon, the 1998 Mars Global Surveyor orbiter had a mass of 450 kg; surface-penetrator probes on current Mars missions have masses as low as 2.5 kg.

The Mars Sojournor rover landed on the martian surface as part of the Pathfinder mission. As noted in the July 1997 issue of *Spaceflight*, this complex and successful spacecraft had a mass of only 12 kg.

A new series of planetary rovers is now under development using 'micro' technology, and may soon be available for missions to Mars, asteroids or other Solar System objects. Laboratory models of these miniaturised spacecraft, as described by N. Amati *et al.*, have masses less than 3 kg and typical dimensions of 30 cm. These experimental micro-rovers can climb 15-degree slopes, traverse 12-cm high obstacles and attain speeds of 15 metres per hour with a power requirement of 2.5 W.

But miniaturisation will surely not end here. Utilising the new science of nanotechnology pioneered by Eric Drexler and others, twenty-first century spaceprobes

Macro Spaceprobes: 1975 timeframe

Example: Mars Viking Lander
Mass 1,000+ kg
Cost about $1 billion

Mini Spaceprobes: 1990s timeframe

Example: Mars Pathfinder Lander
Mass 100 kg
Cost about $100 million

Micro Spaceprobes: 2000+ timeframe

Example: Next Generation
          Planetary Rover
Mass about 10 kg
Cost estimate $10 million

Nano Spaceprobes: 2020(?) timeframe

Example: Dyson's 'Astrochicken'
Mass about 1 kg (?)
Cost estimate $1 million (?)

Fig. 5.1. How spaceprobes shrink as a function of time.

might have a mass 1 kg or less. One of the most interesting (and fanciful) of the nanoprobe proposals is Freeman Dyson's 'Astrochicken'.

Astrochicken would be a 1-kg spacecraft with the capabilities of Voyager. It would be grown rather than built, with a biological organisation and blueprints encoded in DNA. Genetically engineered plant and animal components would be required in Astrochicken. Solar energy would power the craft in a manner analogous (or identical) to photosynthesis in plants. Sensors would connect to Astrochicken's 1-gm computer brain with nerves like those in an animal's nervous system. This space beast might have the agility of a humming bird, with 'wings' that could serve as solar sails, sunlight collectors and planetary-atmosphere aerobrakes. A chemical rocket system for landing and ascending from a planetary surface would be based upon that of the bombadier beetle, which sprays its enemies with a scalding hot liquid jet.

5.2 NANOTECHNOLOGY: THE ART AND SCIENCE OF THE VERY, VERY SMALL

To master nanotechnology and ultimately construct (or grow) something like Astrochicken, it will be necessary to perfect industrial processes capable of manipulating molecular-sized or even atomic-sized objects.

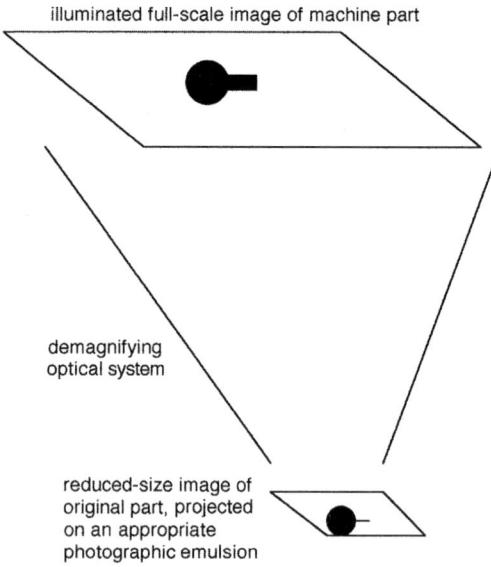

Fig. 5.2. An approach to the manufacture of a nanocomponent.

A number of possible approaches to nanomachining are described in a 1998 paper by R. A. Lawes. One of these, shown in Figure 5.2, is based upon techinques pioneered in the semiconductor industry. A full-scale image of a machine part is demagnified through an optical instrument resembling a microscope used in reverse. The tiny image is projected upon an appropriate light-sensitive substrate. The exposed 'microphotograph' is then developed to obtain the reduced image, and then washed to remove excess photographic emulsion. The emulsion must have appropriate electrical and mechanical properties so that the tiny machine part works in the same manner as its larger model. Micro-sized machines and components have been produced using this approach.

A major limitation to such 'microphotography' is the wavelength of light used. There is no reason why visual-spectral-region optical techniques cannot be further developed for application in lower-wavelength regions of the EM spectrum such as the ultraviolet and even soft X-ray. Since electrons have smaller wavelengths than all but the most energetic photons, electron-optical image demagnifiers should ultimately allow the 'construction' of complex machines composed of individual atoms.

Such a technology will allow construction of really tiny nanospacecraft. In 1998, Anders Hansson reported the results of a NASA Ames Research Center study that considered the design of spacecraft with characteristic dimensions of about 1 μm. Consisting of only a few million atoms, such virus-sized spacecraft could be engineered to travel in a swarm. Their simple computers would be programmed to avoid collision and maintain formation.

Upon arrival at the destination, the swarm of tiny nanoprobes could perhaps link together to form a larger spacecraft. Nanobiology would be applied, as well as nanotechnology. Engineered solar sails or cables could conceivably be replaced by extremely strong and tough biological substances such as spider dragline silk.

Allan Tough has suggested that progress in nanotechnology/nanobiology might be rapid, and the costs of nanoproducts might drop rapidly. By 2021, a human-equivalent computer may cost only $1, and fully developed nanotechnological machines constructed one atom at a time could be marketed by 2050. Nanotechnology might even solve the problem of interstellar communication for an interstellar Astrochicken, which could be launched within the next two centuries. Such a tiny deep-space probe might carry a nano-engineered hyperthin communication antenna to relay information to Earth, or be programmed to construct one from local resources in the destination star system.

It is fortuitous that nanotechnological techniques promise to ultimately reduce the mass of an interstellar probe's communication subsystem. As discussed by J. R. Lesh *et al.*, current technology laser links promise to reduce this mass by a factor of four, since laser beams are better collimated than are radio beams. But even using hyperthin or inflatable beam-focusing optics, the mass of current technology communications subsystems for probes 40 AU or more from the Sun approximates 10 kg.

Exercise 5.2 A major advantage of laser interstellar probe communication links over radio is that lasers (including their microwave variety, 'masers') are well collimated. Radio transmitters, on the other hand, have looser beam collimation; some are omnidirectional. Assume that a space probe is 1 light year from the Earth and transmitting with both radio and laser. Both communication systems generate 10 W of beam power. Assuming that the radio is an omnidirectional, spherically symmetrical transmitter, first calculate the beam irradiance (in $W\,m^{-2}$) reaching the Earth. Do this by dividing beam power by the area of the sphere centred on the transmitter, at the location of the Earth. If the laser has a beam divergence of 0.001 radians, next calculate the laser beam diameter at the Earth (divergence angle in radians is the ratio of beam diameter to transmitter–receiver separation). Calculate laser beam irradiance at the Earth by dividing beam power by beam area at the location of the terrestrial receiver. The receiver area (in square metres) required to gather any beam power (in watts) is the ratio of beam power to beam irradiance (in watts per square metre) at the location of the receiver.

5.3 NANOTECHNOLOGY AND SPACEFLIGHT: NEAR-TERM POSSIBILITIES

In order to implement nanotechnologal advances in an efficient manner, funding agencies often develop 'road maps' to denote significant developmental milestones and the steps necessary to achieve them. A team directed by Al Globus of NASA Ames Research Center has performed this function for near-term application of nanotechnology to aerospace.

Perhaps the first nanotechnological devices to see widespread aerospace application will be tiny computers and very low-mass launch vehicles. The latter would be constructed using nanoengineered thin, diamond-strength materials.

Tiny robots (nanobots) could be used for waste recycling and closed life support applications, as well as in 'smart' spacesuits. Early application of nanobot-engineered structures might be low-mass, large-diameter telescope mirrors; hyperthin solar-sails and solar-cell arrays; and superstrong, hyperthin 'nanotubes' for use in solar-sail cabling and spacecraft tethers. Even a nanotube 'elevator' from the Earth's surface to geosynchronous Earth orbit (GEO) is not beyond reason.

One significant driver for future operations beyond low Earth orbit (LEO) is our desire to explore, divert and possibly mine asteroids and comets that approach the Earth. One way to eventually apply nanotechnology to asteroid mining, as discussed by T. McKendrie, is the 'asteroid assembler'. This would be a nanodevice that would land on an appropriate asteroid or comet nucleus, grow larger and reproduce by planting appropriate 'seeds' in the object's material, and ultimately begin to mine the asteroid or comet. [McKendrie, T., 'Design Considerations for Carbonaceous Asteroid Assemblers', *Journal of the British Interplanetary Society*, **51**, 153–160 (1998).]

A very interesting nanoproduct that is now undergoing widespread study is the 'Fullerene nanotube', named after visionary architect Buckminster Fuller. As discussed by D. Brenner *et al.*, Fullerine nanotubes are essentially single sheets of graphite wrapped into tubes 10 nm across. These could be adopted as cables or in sensor design, and are characterised by very high strength and low specific gravity.

5.4 NANOTECHNOLOGY AND SPACEFLIGHT: LONG-TERM POSSIBILITIES

Nanotubes and other tiny structures could also influence the design of huge solar-sail starships capable of carrying large human populations to α Centauri within a 1,000-year flight time. Matloff has considered the design advantages of solar-sail nano-cables much stronger than diamond or constructed using highly reflective, thin and strong 'nanoribbons'. From equation (4.22), solar-sail cable mass increases with payload mass and decreases with increasing cable tensile strength. Reflective thin cables allow stronger and higher accelerations or payloads because solar radiation pressure on the cable partially offsets inertial forces.

But perhaps a more elegant approach to interstellar colonisation is to team nanobiology with nanotechnology to shrink the colonists and therefore the spacecraft. In 1996, Anders Hansson suggested that tiny, living spacecraft might be applied to the interstellar expansion of terrestrial life and civilization.

Imagine a living Astrochicken with miniaturised propulsion subsystems, autonomous computerised navigation via pulsar signals, and a laser communication link with Earth. The craft would be a bioengineered organism. After an interstellar crossing, such a living Astrochicken would establish orbit around a habitable planet. The ship (or being) could grow an incubator/nursery using resources of the target solar system, and breed the first generation of human colonists using human eggs and sperm in cryogenic storage.

5.5 POSSIBLE LIMITS TO NANOTECHNOLOGY

In the early phases of any new technology, the enthusiasm of its adherents often overcomes consideration of its limitations in practice. This is certainly true of nanotechnology. Perhaps as well as considering its potential for future development, we might also consider the factors that in practise might limit nanotechnology for both future terrestrial and extraterrestrial applications.

Consider, for example, the popular concept of a nanobot population injected into the human bloodstream to kill cancer cells, reduce cholesterol, and so on. Such nanobots might consist of miniaturised robotic arms of the type considered by Salvatore Santoli, combined with nanocomputers and sophisticated sensor systems.

Such a device, if of molecular size, seems suspiciously like 'Maxwell's demon'. If possible, such a tiny entity could separate hot and cold molecules in a glass of water. A fleet of nanobots could perhaps convert a lukewarm glass of water into a fluid that is boiling on its left side and freezing on its right.

This is probably ruled out by the Second Law of Thermodynamics, since each nanobot would generate a certain amount of waste heat. Even terminally ill patients might like to know what nanobot-generated waste heat would do to their bloodstreams and immune systems before the injection of a billion-nanobot fleet. (On the other hand, we should perhaps not be too hasty. In his 1999 *Journal of the British Interplanetary Society* paper, Santoli informs us that thermodynamics will work differently at the nanolevel than at the macro-level.)

Fast, tiny starships might also have some developmental limitations. In the Autumn 1999 issue of *SearchLites* (a publication of the SETI league in Little Ferry, New Jersey) some problems affecting tiny-starship operation are debated by Paul Davies, Allen Tough and Mario Zadnik. One limitation to the 'nanostarship' concept is Galactic cosmic rays. These highly energetic (million-billion eV) atomic nuclei have a Galactic flux of a few particles per square centimetre per second. Because of their penetration depth (up to a one metre), nano-machined tiny starships might be damaged or incapacitated by cosmic rays soon after launch.

But tiny living organisms have survived in space for periods measured in years. Terrestrial bacteria survived on the lunar surface for years on the Surveyor robot arm returned to Earth by the Apollo 12 crew in 1969. Since life forms – even bacteria – are self-repairing nanomachines, nanostarships could protect against cosmic-ray damage by using nanobiological techniques to incorporate bacterial self-repair mechanisms.

Whatever its ultimate limitations, nanotechnology/nanobiology promises to revolutionise life on Earth and concepts of life's expansion beyond the terrestrial biosphere. These technologies should be routinely monitored by space-travel enthusiasts for new applications and breakthroughs. It would be interesting to visit the nanotechnological facilities of the twenty-second century to learn which predictions of contemporary nanotechnologists have borne fruit.

5.6 BIBLIOGRAPHY

Amati, N., Chiaberge, M., Genta, G., Miranda, E. and Reyneri, L. M., 'Twin Rigid-Frames Walking Microrovers: A Perspective for Miniaturisation', *Journal of the British Interplanetary Society*, **52**, 301–304 (1999).

Brenner, D., Scholl, J. D., Mewkill, J. P., Shenderova, O. A. and Sinnott, S. B., 'Virtual Design and Analysis of Nanometre-Scale Sensor and Device Components', *Journal of the British Interplanetary Society*, **51**, 137–144 (1998).

Burnham, D. and Salmon, A., 'A Decade of Mars Missions', *Spaceflight*, **38**, 400–403 (1996).

Eric Drexler, K., *Engines of Creation: The Coming Era of Nanotechnology*, Anchor/Doubleday, New York (1986).

Dyson, F., *Infinite in All Directions*, Harper & Row, New York (1985).

Globus, A., Bailey, D., Han, J., Jaffe, R., Levit, C., Merkle, R. and Srivastava, D., 'Aerospace Application of Molecular Nanotechnology', *Journal of the British Interplanetary Society*, **51**, 145–152 (1998).

Hansson, A., 'Towards Living Spacecraft', *Journal of the British Interplanetary Society*, **49**, 387–390 (1996).

Hansson, A., 'From Microsystems to Nanosystems', *Journal of the British Interplanetary Society*, **51**, 123–126 (1998).

Lawes, R. A., 'Microsystems and How to Access the Technology', *Journal of the British Interplanetary Society*, **51**, 127–132 (1998).

Lesh, J. R., Ruggier, C. J. and Casarone, R. J., 'Space Communications Technologies for Interstellar Missions', *Journal of the British Interplanetary Society*, **49**, 7–14 (1986).

Matloff, G. L., 'The Impact of Nanotechnology upon Interstellar Solar Sailing and SETI', *Journal of the British Interplanetary Society*, **49**, 307–312 (1996). Also see G. L. Matloff and B. N. Cassenti, 'Interstellar Solar Sailing: the Effect of Cable Radiation Pressure', IAA 4.1-93-709.

Mileikowsky, C., 'How and When Could We Be ready to Send a 1000-kg Research Probe With a Coasting speed of 0.3 c to a Star?', *Journal of the British Interplanetary Society*, **49**, 335–344 (1996).

Santoli, S. 'Enroute to Basic Machine Members for Molecular Manufacturing: Hypothesis for Rigid-Link Nanorobotic Arms', *Journal of the British Interplanetary Society*, **45**, 427–432 (1992).

Santoli, S., 'Developing the Concept of a Rigid-Link Nanorobotic Arm Working in a Dense Fluid', *Journal of the British Interplanetary Society*, **47**, 341–342 (1994).

Santoli, S., 'Molecular Nanotechnology – Rethinking Chemistry, Computation and Interstellar Flight', *Journal of the British Interplanetary Society*, **52**, 336–342 (1999).

Tough, A., 'Small, Smart Interstellar Probe', *Journal of the British Interplanetary Society*, **51**, 167–174 (1998).

Yenne, W., *The Encyclopedia of US Spacecraft*, Exeter, New York (1985).

6

The nuclear option

> *The dream of flight was one of the noblest and one of the most disinterested of all man's aspirations. Yet it led in the end to that B-29 driving in passionless beauty toward the city whose name it was to sear on the conscience of the world.*
>
> Arthur C. Clarke, *The Promise of Space* (1968)

Noted space visionary Arthur C. Clarke succeeds, in the short quote above, in succintly summarising the quandary of those who seek to apply nuclear-fission technology to space travel. Even though their hearts may be pure and their goals justified, the simple fact that their methods were used first to kill hundreds of thousands of humans at Hiroshima and Nagasaki and later to fuel the long mock peace of the Cold War, insures a vast reservoir of public distrust directed towards the nuclear scientists.

In the post-Cold War era, many usually rational citizens are afflicted with an attitude referred to as 'NIMBY' – 'not in by back yard'. This is why it is difficult to find a safe, underground repository for US nuclear waste, and why Florida residents' fears of a very unlikely launch disaster threatened the 1997 launch of the Cassini probe to Saturn. The probe is powered by a small radioisotope thermal generator.

But public attitudes have changed before, and may change again. It would seem irrational to discard the nuclear-spaceflight option after so many decades of development effort. As mentioned in previous chapters, certain methods of space exploration may be possible only through the use of nuclear power. And if Earth is threatened by an approaching asteroid on a collision course, the nuclear option may be our only option.

6.1 NUCLEAR BASICS

Although their goals may be different, the operation of all nuclear reactors, bombs and rockets is rooted in the same technique: the conversion of mass (m) into energy (E), according to Einstein's famous formula $E = mc^2$, where c is the speed of light.

We consider here three basic types of nuclear reaction with spaceflight application: fission, fusion, and matter/antimatter annihilation.

Fission – the splitting of massive atomic nuclei such as uranium-235 into less massive atoms and energy – was first suggested by German scientists on the eve of World War II, and applied late in that conflict to the development of the first atomic reactors and bombs during the American Manhattan Project. To date, all power reactors used to generate electricity are fission devices, the triggering mechanisms (at least) of all nuclear weapons are fission based, and those nuclear rockets that could soon become operational are fission rockets. Many radioisotopes used in hospitals and industry are produced in fission reactors; but sadly, all nuclear waste generated by our species to date can also be attributed to fission.

Nuclear fusion – in which low-mass nuclei are joined to form more massive nuclei and energy – is the power source of the Sun and billions of other main-sequence stars. Our technology is not yet quite able to fuse light nuclei in power reactors and obtain net energy. Even our hydrogen bombs are triggered by smaller fission 'devices'. While a star sustains fusion reactions deep in its interior through the pressure of its self-gravitation, human researchers apply electromagnetic-field compression techniques or blast fusion micropellets with laser or electron beams. Very controversial fusion possibilities, often listed under the rubric 'cold fusion', are considered in a following chapter. Although ultimate 'hot' fusion reactors may be relatively free of radioactivity, those reactions applicable in the early twenty-first century will produce some nuclear waste, although not as much as fission reactors produce.

Both fission and fusion convert only a tiny fraction (less than 1%) of their total mass to energy. From the point of view of nuclear rocketry, the antimatter/matter annihilation reaction is superior because 100% of the reactant mass is converted to energy, and much of this energy can be directly channelled into rocket exhaust. But antimatter (charge-reversed matter) does not exist in nature during the Universe's present epoch, and is extremely expensive to produce in useful quantities. Because an antiproton and a proton will attract each other by Coulomb's law and instantly annihilate upon contact, long-term storage of antimatter is challenging. Although matter/antimatter annihilation (or simply 'antimatter') rockets can theoretically attain high relativistic velocities, many researchers more conservatively plan to utilise tiny amounts of this resource to initiate fusion in much larger fusion 'micropellets'.

Figure 6.1 schematically illustrates the operation of any nuclear rocket. Nuclear fuel is injected into the reactor at a controlled rate, and is reacted to produce energy. Some of the energy may be included in a reactor exhaust stream, and some of the energy may be transferred to a nuclear-inert reaction fuel. Because of the limitations of thermodynamics, it will always be necessary to radiate a fraction of the nuclear energy generated by the reactor as waste heat.

If dM_{nf} and dM_{if} are the nuclear and inert fuel masses ejected during an infinitesimal time interval dt, M_s is the total spacecraft mass at the start of dt, and V_s is the ship velocity at the start of dt, conservation of linear momentum can be applied to the spacecraft and the two exhaust streams:

$$M_s V_s = (M_s - dM_{nf} - dM_{if})(V_s + dV_s) + dM_{nf}(V_s - V_{e,nf}) + dM_{if}(V_s - V_{e,if}) \quad (6.1)$$

Nuclear basics

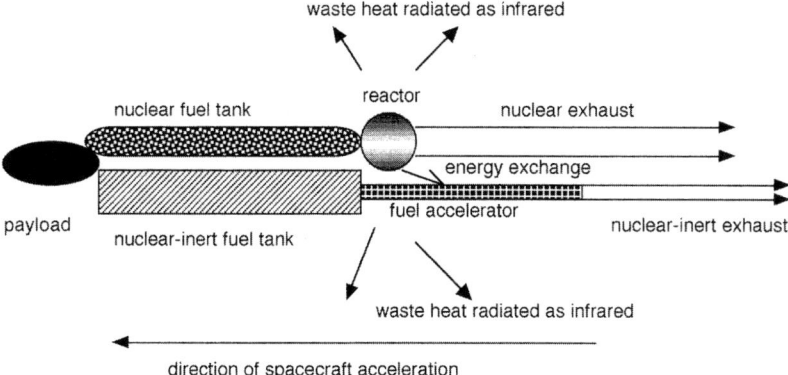

Fig. 6.1. Schematic operation of a generalised nuclear rocket.

where $V_{e,nf}$ and $V_{e,if}$ are respectively the exhaust velocities of nuclear and nuclear-inert fuel streams, and dV_s is the ship's increase in velocity during time interval dt.

Although nuclear-thermal rockets, in which nuclear energy is used to directly heat the nuclear-inert fuel, are certainly possible for interplanetary applications, we instead consider here nuclear rockets in which some of the nuclear energy is first converted into electrical energy, which is then transferred to the kinetic energy (KE) of the nuclear-inert exhaust. If the mass/energy conversion efficiency of the nuclear reaction is Φ_{nf}, ε_{nf} is the fraction of released nuclear energy that is transferred to the nuclear-exhaust stream, ε_{ne} is the efficiency of nuclear-energy to electric conversion, and ε_{ei} is the efficiency at which this electrical energy is converted into nuclear-inert exhaust kinetic energy, we can calculate exhaust kinetic energies, waste heat radiated (WHR) and exhaust velocities using Einstein's basic mass/energy conversion equation:

$$dKE_{nf} = \Phi_{nf}\varepsilon_{nf}dM_{nf}c^2 = (\tfrac{1}{2})dM_{nf}V_{e,nf}^2 \tag{6.2a}$$

$$dKE_{if} = \Phi_{nf}\varepsilon_{ei}\varepsilon_{ne}dM_{nf}c^2 = (\tfrac{1}{2})dM_{if}V_{e,if}^2 \tag{6.2b}$$

$$WHR = (1 - \varepsilon_{nf} - \varepsilon_{ei}\varepsilon_{ne})\Phi_{nr}dM_{nf}c^2 \tag{6.2c}$$

Equations (6.2) can be manipulated to yield simple equations for nuclear and nuclear-inert exhaust velocities:

$$V_{e,nf} = (2\Phi_{nf}\varepsilon_{nf})^{1/2}c \tag{6.3a}$$

$$V_{e,if} = \left[2\Phi_{nf}\varepsilon_{ei}\varepsilon_{ne}\left(\frac{dM_{nf}}{dM_{if}}\right)\right]^{1/2}c \tag{6.3b}$$

Table 6.1 lists representative values (from the literature) of the mass/energy conversion efficiency of the nuclear reaction, Φ_{nf}, for various nuclear reactions.

Table 6.1. Values of mass–energy conversion efficiency (Φ_{nf}) for fission, fusion and matter/antimatter annihilation reactions

Reaction	Mass–energy conversion efficiency	Literature source
Fission	0.00075	Shepherd (1999)
Fusion	0.004	Shepherd (1999)
Antimatter	1.0	Forward and Davis (1988)
,,	0.6 (useful energy)	Vulpetti (1986)

In the following discussion, values of various energy-transfer and energy-utilisation factors in equations (6.3a) and (6.3b) are given for various existing and hypothetical nuclear space propulsion systems.

If we substitute these values of mass/energy conversion efficiency into equation (6.3a) and assume that all released nuclear energy is transferred to the exhaust stream ($\varepsilon_{nf} = 1$), exhaust velocities for ideal nuclear rockets can easily be calculated.

Exercise 6.1 Calculate exhaust velocities for ideal fission, fusion, and matter/antimatter annihilation rockets as fractions of the speed of light and in km s^{-1}.

One factor that determines the fraction of released nuclear energy that can be productively utilised in a real (or non-ideal) nuclear rocket is the form taken by that energy. Generally, nuclear reactions deposit only a fraction of their energy in the kinetic energy of electrically charged reaction products. This is the energy fraction that can be directly utilised for propulsion with high efficiency. But some of the energy released in the nuclear reaction may appear instead in the form of electrically neutral neutron kinetic energy, X-rays or gamma rays, or neutrinos (zero- or low-mass particles that are extremely nonreactive with matter).

In 1970, Matloff and Chiu considered a 'two-stream' nuclear rocket of the form pictured in Figure 6.1, and described by equations (6.2) and (6.3). Two different ways of utilising the portion of nuclear energy not emitted as charged-particle kinetic energy were compared. One of them is the nuclear–electric drive (NEP), in which some of the non-charged-particle energy is used to accelerate a nuclear-inert ion fuel. The second method is a variety of the 'photon drive' in which some released energy in the form of X-rays, gamma-rays and neutrons is used to pump an efficient gas laser that emits its beam as exhaust. Because ions have more momentum than do photons, the NEP is more effective unless a high percentage of the nuclear fuel mass is converted to energy.

Most existing or proposed nuclear propulsion systems are 'one-stream' only. Exhaust is either in the form of nuclear reaction products or accelerated inert propellant, But, as presented in a following chapter, the two-stream propulsion concept surfaces once again as the ram-augmented interstellar rocket (RAIR) – a variant of the interstellar ramjet.

6.2 NUCLEAR-ELECTRIC PROPULSION (NEP)

One form of fission propulsion that might see application in early robotic interstellar exploration missions is the nuclear-electric, or nuclear-ion drive (NEP). Operation of a typical NEP design is illustrated in Figure 6.2. In this approach, nuclear fuel is consumed but not exhausted. Some of the released nuclear energy is transferred to an electrical (or magnetic) thruster that accelerates ions of nuclear-inert fuel as rocket exhaust. Ion fuels are usually materials like caesium, argon or mercury that are easy to ionise and chemically relatively non-reactive. Ion fuel is usually exhausted at a very slow rate over a long period of time. Although thrust is low (typically 10^{-4} g, where g is one Earth surface gravity), high spacecraft velocities can build up over a period of weeks or months. With exhaust velocities in the neighbourhood of 100 km s^{-1}, NEP is a candidate propulsion system for some of the early extrasolar probes.

Equation (6.3b) defines NEP exhaust velocity. If we define a constant $Q_{nep} = dM_{nf}/dM_{if}$, the ratio of nuclear-fuel to nuclear-inert fuel consumed in time interval dt, equation (6.3b) can be substituted into equation (6.1) for the case of no nuclear exhaust. Rearranging and performing the substitution $dM_s = dM_{if}$, since ship mass decreases as ion fuel is exhausted, we obtain

$$\frac{\Delta V_s}{V_{e,if}} = -\int_{M_f+M_0}^{M_0} \frac{dM_s}{M_s} \tag{6.4}$$

where ΔV_s is the total increase in spacecraft velocity, M_f is the total ion fuel mass, and M_0 is the ship's empty (unfuelled) mass. Defining the mass ratio M.R. as the ratio of $(M_f + M_0)$ to M_0, and substituting for ion exhaust velocity, equation (6.4) can be solved to obtain

$$M.R. = \exp\left[\frac{\Delta V_s}{(2\Phi_{nf}\varepsilon_{ei}\varepsilon_{nf}Q_{nep})^{1/2}c}\right] \tag{6.5}$$

Exercise 6.2 Verify all steps in the derivation of equation (6.5).

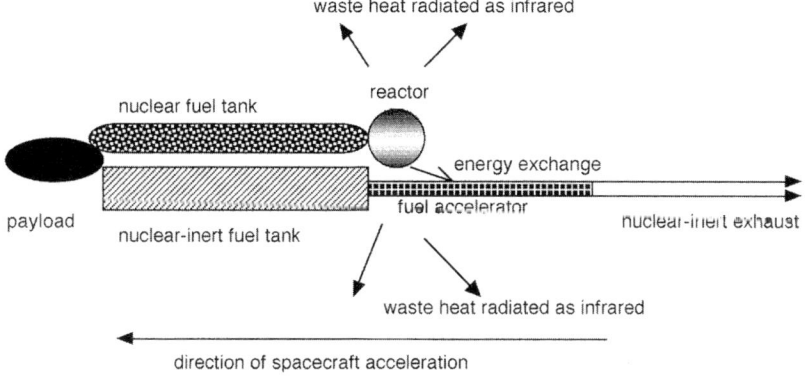

Fig. 6.2. Schematic operation of a nuclear-electric rocket.

We can apply Table 6.1 and equations (6.3b) and (6.5) to evaluate performance of NEP for an interstellar precursor mission based on modern NEP technology. From Table 6.1, $\Phi_{nf} = 0.00075$ for the fission reaction. From Ronald Lipinski et al., the fraction of fission energy that can be transferred to the ion exhaust ($\varepsilon_{ei} \varepsilon_{ne}$) is 0.5–0.8; we will adapt 0.65 as a median value. Also from Lipinski et al., the ratio of nuclear fuel burned to ion mass exhausted (Q_{nep}) is about 0.0005. Substitution into equation (6.3b) reveals that an ion-exhaust velocity of about $0.0007 c$ (200 km s^{-1}) is possible for this configuration. David Fearn has argued that much higher exhaust velocities may be possible for an 'ultimate' NEP.

Exercise 6.3 Insert the value 200 km s^{-1} for ion-exhaust velocity into equation (6.5). What mass ratio is required to accelerate an NEP spacecraft to terminal velocities of 50 and 100 km s^{-1}?

One limitation for NEP applied to true interstellar flight is the low specific power. From Shepherd's recent contribution, 50 kW kg^{-1} is the best we might reasonably hope for. Let us say, for example, that an NEP reactor is charged with 10 kg of uranium fuel, all of which is expanded during a 1-year acceleration period. The nuclear fuel exhaust rate is therefore $3.2 \times 10^{-7} \text{ kg s}^{-1}$. Applying Einstein's mass/energy equation with the correct fission efficiency from Table 6.1, we find that the onboard reactor generates about $2 \times 10^4 \text{ kW}$, and the reactor mass is about 400 kg. For very high performance missions, the reactor mass will dominate the mass budget.

One analyst, Graeme Aston, has attempted to estimate from current technology what would be optimistically possible if NEP technology were developed to its limits. In Aston's parameterisation, reactor mass is given:

$$M_{nep,react} = 45.8 \left[\frac{P_{nep}}{\varepsilon_{ne} \varepsilon_{ei}} \right]^{0.36} \text{ kg} \quad (6.6)$$

where P_{nep} is the reactor power in kilowatts and the efficiency factors in the denominator (which represent the total efficiency) are defined above.

Assuming, with Aston, that the reactor is a rotating-bed reactor fuelled with uranium-233, the total mass of fission fuel is given by:

$$M_{nep,nf} = 4.25 \times 10^{-4} \left[\frac{P_{nep}}{\varepsilon_{ne} \varepsilon_{ei}} t_{burn,yr} \right] \text{ kg} \quad (6.7)$$

where $t_{burn,yr}$ is the NEP engine-burn time in years.

The ion beam energy, in electron volts, is a function of the square of the ion exhaust velocity in m s^{-1}, which is defined above:

$$E_{nep,beam} = 1.05 \times 10^{-6} V_{e,if}^2 \text{ eV} \quad (6.8)$$

Aston next defines ion beam current as

$$J_{nep,beam} = 4.78 \times 10^4 \frac{Th_{nep}}{0.01 V_{e,if}} \text{ amperes} \quad (6.9)$$

where Th_{nep} is the ion engine thrust in Newtons. The engine power in kW, P_{nep} in equation (6.7), is 0.001 $(E_{nep,beam})(J_{nep,beam})$.

From the results of equations (6.8) and (6.9), Aston next estimated the masses of various ion engine subsystems. He assumed a (mercury) propellant-utilisation efficiency of 0.98, and a power-conditioning efficiency of 0.95. A high-temperature Brayton-cycle energy conversion system with an efficiency of 0.4–0.45 was assumed to convert nuclear energy into electrical energy. The mass of the low-power beam-current generating equipment (ion source and pre-accelerator subsystem low-power conditioning equipment) is given as

$$M_{nep,lp} = N_{nep}(10 J_{nep,beam}^{0.5} + 2.6 J_{nep,beam}) \text{ kg} \qquad (6.10)$$

where N_{nep} is the number of NEP engines. The mass of the NEP high-power power conditioning equipment consists of a fixed-mass 5,000-kg accelerator and a very high-power Cockroft–Walton power supply with a mass of 0.009 (P_{nep}) kg. For reactor power levels above 1,000 mW, mass of the energy-conversion system (including compressor pumps and turbo alternator) is 0.02 (P_{nep}) kg.

The final probe power-system mass to be estimated by Aston was that of the radiator necessary to radiate 55–60% of the energy generated by the reactor to space. He assumed a moving-belt radiator with a mass

$$M_{nep,rad} = 0.012 \left(\frac{1 - \varepsilon_{ne} \varepsilon_{ei}}{\varepsilon_{ne} \varepsilon_{ei}} \right) P_{nep} \text{ kg} \qquad (6.11)$$

Aston calculated the performance of an NEP starprobe with a payload mass of 5,000 kg, an additional 5% of initial mass for structure, an ion-exhaust velocity of 4,000 km s^{-1}, and an engine burn time of 64.9 years. The spacecraft terminal velocity was 0.012c, the total reactor power was 1 mW, and the total energy-conversion efficiency was 0.4. With 270,000 kg of mercury ion propellant, 113,000 kg of nuclear fuel, a 1,600-kg reactor mass, and about 45,000 kg allocated to electric-engine components, the total flight time to α Centauri for this system was calculated as 389 years.

In 1992, R. A. Bond and Anthony Martin presented performance estimates for advanced, technologically feasible NEP spacecraft. They concluded that maximum power-conversion efficiencies of 0.2 will be possible, and obtained thruster masses about three times greater than those derived from Aston's formulae. They suspect that Aston's reactor-mass estimates are very optimistic and may never be achieved.

Nuclear-electric propulsion systems are massive. Nanotechnological reduction in payload mass has little effect on total mission mass requirements. But if we must launch a probe to land on and explore an object in the Kuiper Belt or Oort Cloud, the NEP is our best propulsion option for the probe-deceleration phase. Contrary to popular opinion, the NEP need not pose a launch hazard if the reactor is launched in 'cold' mode and not turned on until the spacecraft has escaped the Earth.

One interesting fission-type device that has been considered recently by Robert J. Noble is radioisotope-electric propulsion, in which mission power is obtained from the decay of radioactive isotopes. Unlike NEP, this system can be miniaturised. But it will present environmental risks, since the powerplant cannot be launched 'cold'.

6.3 NUCLEAR-PULSE PROPULSION: ORION, DAEDALUS AND MEDUSA

Nuclear-electric propulsion may never be capable of true interstellar flight. However, the same cannot be said for nuclear-pulse propulsion, which is illustrated in Figure 6.3.

Project Orion – probably the first serious nuclear-pulse proposal – began its life in the early Space Age as a secret USAF project. It was later incorporated in NASA plans for the first Moon-landing missions, in case the Saturn V failed. Before its cancellation in the mid-1960s, several Orion prototypes were tested in the Earth's

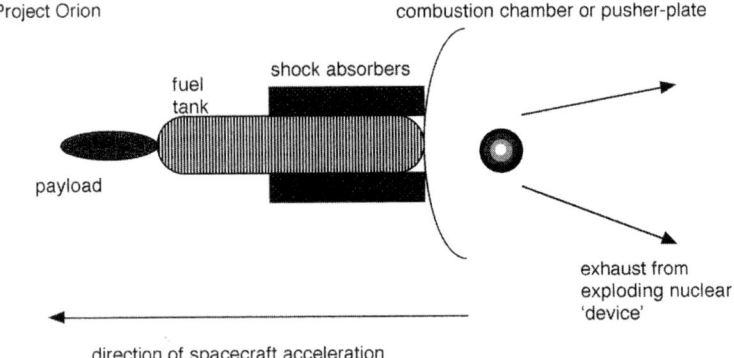

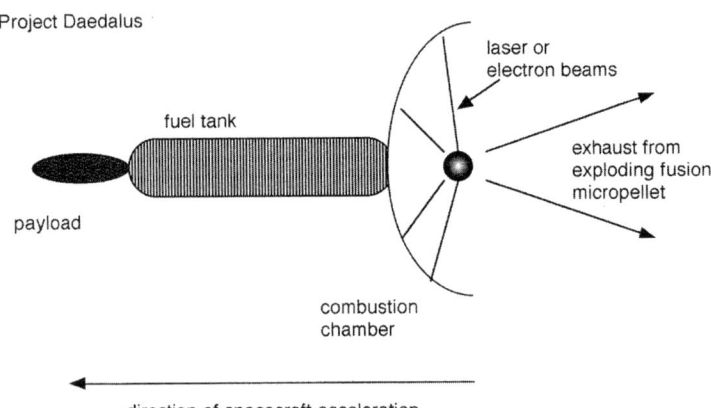

Fig. 6.3. Two approaches to nuclear-pulse propulsion.

atmosphere using chemical (not nuclear) charges. An Orion prototype is on display in the Smithsonian Air and Space Museum, and a photographic sequence of an Orion prototype in flight appears in *The Starflight Handbook*. After the project's cancellation, Freeman Dyson published (in 1968) a design for an interstellar Orion.

It is perhaps just as well that the original Earth-launched Orion was never constructed, because it would have been an environmental nightmare. Nuclear-pulse propulsion (NPP) should be used only in space, hopefully far from any inhabited planet.

In the basic Orion concept, the payload and fuel tank are separated from a combustion chamber (or pusher plate) by enormous shock-absorbers. The reason for this separation (which from the crew's point of view should be as large as possible) is the nature of the fuel. This is not simply a nuclear reactor; the fuel consists of nuclear bombs (often called 'devices' to sanitise the concept). To prevent its evaporation, the pusher plate would be coated with an ablative material. Unless shaped nuclear charges are applied, no more than half the bomb-exhaust actually interacts with the pusher plate or combustion chamber.

With kiloton-sized fission devices, a Project Orion vehicle launched from Earth might have an exhaust velocity as high as $200\,\mathrm{km\,s^{-1}}$, and much greater thrust than an NEP-propelled spacecraft. The Orion pusher plate assembly would also be considerably less massive than an NEP reactor.

With Dyson's interstellar Orion, the spacecraft is much larger and can operate only in space. The kiloton-sized fission devices are replaced by 1-megaton hydrogen bombs. These operate by the fusion of deuterium atoms, which are heated by energy released from a small fission 'trigger'.

Without violating national security protocols, Dyson was unable to accurately predict performance of his interstellar Orion. He did, however, conclude that for a mass ratio of 4, if fuel is used for both acceleration and deceleration, the terminal (interstellar cruise) velocity of a thermonuclear Orion is between $0.0035c$ and $0.035c$. If all of the hydrogen bombs in humanity's arsenals were devoted to the task, small populations could be transferred to α Centauri on voyages of 100–1,000-year duration. What a lovely use for the bombs!

Exercise 6.4 Calculate the exhaust velocity range for Dyson's interstellar thermonuclear Orion. First double the interstellar cruise velocity to obtain the total velocity increment (since the spacecraft must both accelerate and decelerate); then apply equation (4.4) to determine the range of exhaust velocities. From the value for the fusion mass/energy conversion fraction in Table 6.1 and equation (6.3a), calculate the range for ε_{nf}, the fraction of nuclear energy transferred to the kinetic energy of the nuclear exhaust stream (also called the burn fraction).

In an effort to reduce the radioactive emissions from an Orion spacecraft, F. Winterberg has investigated the possibility of triggering a thermonuclear explosion by means other than a fission device. In 1977 he considered igniting a deuterium–tritium fusion reaction with chemical explosives. In an earlier 1971 contribution he had considered another mode of igniting a fusion reaction – intense, relativistic electron beams. This earlier contribution led directly to Project Daedalus, a starship study conducted by the British Interplanetary Society during the 1970s and 1980s (Alan Bond et al., 1978).

Instead of megaton-sized H-bombs, Daedalus uses for fuel micropellets of fusable isotopes. The most easily ignitable fusion-fuel combination is a mixture of deuterium and tritium, each of which are heavy isotopes of hydrogen. This fuel combination was rejected by the Daedalus design team because much of the fusion energy is released in the form of thermal neutrons, which result in nuclear radiation after absorption in the combustion chamber walls.

The Project Daedalus fuel combination of choice was a mixture of deuterium and helium-3, a light isotope of helium. Deuterium is rather common in nature, and this isotope mix produces many fewer neutrons than the deuterium–tritium mix. A drawback is the extreme rarity of helium-3 in the terrestrial environment.

Helium-3 is, however, found in the solar wind and atmospheres of the giant planets, and Daedalus designers considered a number of options to obtain the millions of kilogrammes required for a full-scale starprobe or starship. These included strip-mining the upper layers of lunar soil (where the solar wind has deposited small concentrations of this isotope), producing helium-3 on Earth, mining the solar wind directly, and mining the atmospheres of the outer planets.

The option selected was to insert robotic helium-mining packages below balloons suspended in the jovian atmosphere. Periodically, the separated helium isotope would be rocketed to fuel-processing stations orbiting Jupiter. Obtaining fuel for Daedalus would indeed be a major task!

The Orion pusher plate would be replaced by a combustion chamber in which electromagnetic fields would reflect the high-velocity charged-particle exhaust from the fusion reaction. Calculations revealed that the exhaust velocity of a Daedalus probe would be as high as $0.03\,c$ (about $9{,}000\,\text{km}\,\text{s}^{-1}$).

Exercise 6.5 Assuming an exhaust velocity of $0.03\,c$, estimate the effective burn fraction (ε_{nf}) for a Daedalus probe. For an unfuelled spacecraft mass of 10^6 kg and an (undecelerated) interstellar cruise velocity of $0.1\,c$, how much fusion fuel is required? If 50% of the fuel mass is helium-3, how much of this rare isotope is required to propel this interstellar spacecraft?

As Hyde et al. discussed in 1972, fusion micropellets can be compressed and ignited by electron beams. Such 'inertial' fusion reactors are already in operation in defence laboratories, simulating (on a small scale) thermonuclear explosions. The Daedalus team instead adopted Winterberg's suggestion of electron-beam ignition.

A number of Daedalus follow-on concepts exist. We might consider lithium-proton or boron-proton fusion reactions. Although more difficult to ignite, these reactions use very common reactants and are aneutronic. In 1977, Winterberg suggested staged microexplosions in which a small helium-3–deuterium pellet ignites a larger boron-proton or lithium-proton pellet after it is ignited by electron or laser beams. (See M. L. Shmatov (2000) for a review of staged fusion microexplosion literature.)

But like Orion, Daedalus would be huge. Both a 10-kg and a 100,000-kg payload would probably require spacecraft with unfuelled masses in the multimillion-kilogramme range. A surprising concept which might greatly reduce spacecraft mass is Johndale Solem's 'Medusa', which combines elements of nuclear-pulse and solar-sail propulsion. Medusa replaces the massive combustion chamber/pusher plate with a gossamer canopy joined to the payload by high-tensile strength cables. Calculations revealed that if the canopy is sufficiently strong and radiation-resistant, it can withstand the nearby ignition of micropellets or even small nuclear devices. If a low-mass field generator could be implaced in this canopy, charged-particle Daedalus exhaust can be reflected with little or no canopy degradation.

Medusa would move through space in a manner analogous to a jellyfish's transit through the ocean – thus the inspiration for the name. Although much analysis remains to be carried out, the Medusa concept might allow great reduction in the mass of a nuclear-pulse starprobe.

6.4 INERTIAL ELECTROSTATIC CONFINEMENT AND GAS-DYNAMIC MIRROR FUSION

Many field geometries have been considered for the confinement of fusion plasmas. One of these – inertial electrostatic confinement (ICE) (Bussard (1991) and Miley *et al.* (1997 and 1999)) – is under serious consideration at the NASA Marshall Space Flight Center, because of its potential space-propulsion application.

Figure 6.4 shows the configuration of an ICE reactor. An electric field is arranged around the interior of a spherical chamber so that a fusion plasma is contained between the walls of the chamber and a central grid. Fusion takes place close to the central grid. As shown in the figure, electrical energy from the fusing plasma

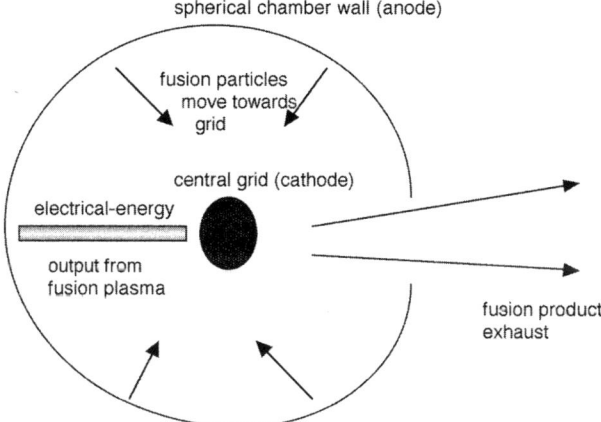

Fig. 6.4. An inertial electrostatic confinement fusion reactor and its application to space propulsion.

could be obtained from the central region. If the high-voltage outer 'anode' near the spherical chamber walls is slightly leaky, fusion-heated exhaust can be expelled out of the rear of a space vehicle. Such a reactor has great promise for the ram-augmented interstellar rocket (RAIR, described in a following chapter) and other 'two-stream' rocket concepts.

An alternative fusion geometry with possible space-propulsion applications has been pioneered by T. Kammash and M.-J. Lee. Based upon the gas-dynamic mirror (GDM) concept, this approach is theoretically capable of fusing tritium and deuterium to obtain a specific impulse of at least 200,000 seconds, and thrusts in the kiloNewton range.

In a GDM reactor, magnetic mirrors are used to confine an injected deuterium–tritium plasma for a sufficiently long time for fusion to occur. Magnetic confinement is stronger at the ends of the cylindrical chamber than at the middle, which prevents most of the plasma from escaping through the ends.

The plasma density in a GDM machine is high enough for the mean free path of the reacting ions to be much shorter than the machine's dimensions. This causes the fusion-heated plasma to act hydrodynamically like a gas heated in a reaction chamber, with a hole allowing hot gas to escape into a vacuum. Fusion products from a properly configured GDM machine should be expelled out of the rear as rocket exhaust.

Another fusion reactor geometry of interest is the 'colliding beam' approach of Norman Rostoker *et al.* (1997). If this machine proves practicable, aneutronic fusion might be possible between colliding beams of protons and boron-11.

6.5 ANTIMATTER: THE ULTIMATE FUEL

Perhaps because of television's *Star Trek* series, everyone has heard of antimatter. Unlike fission and fusion, which convert less than 1% of the reactant mass to energy, all of the mass of a matter/antimatter reaction is converted into energy.

The matter/antimatter reaction violates no physical laws, and the technology of antimatter production and storage is making great strides. The only obstacle to the construction of large antimatter-fuelled rockets capable of approaching the speed of light is the enormous cost of this resource.

Forward and Davis outline the early history of antimatter research. After its prediction by Paul Dirac in 1929, the antielectron (positron) was discovered – using cosmic-ray measuring plates – by Carl Anderson in 1932. Discovery of the more massive antiproton required an energetic nuclear accelerator, and was accomplished by a group directed by Emilio Segre in 1955.

A particle and its antiparticle have opposite electrical charge and therefore attract one another. The mass of both interacting particles is converted into gamma rays. Eugen Sanger applied this effect to his annihilation photon rocket concept in 1965. An annihilation photon rocket operates by first combining matter with equal amounts of antimatter, and then reflecting the gamma rays out the ship's rear as exhaust. Unfortunately, the demands of such gamma-ray reflection seem somewhat beyond our technology.

Inspired, perhaps, by Robert Forward, a small band of interstellar-flight researchers began to study antimatter in the early 1980s. In 1982, P. F. Massier reviewed early concepts for antimatter production and long-term storage. Existing antimatter 'factories' are very inefficient, as they operate by projecting a very energetic particle beam against a stationary metal target. A few antiparticles are produced through the beam/target interaction, and a small fraction of these are collected.

An improved and uprated solar-powered antimatter-production facility was proposed by G. Chapline in 1982. The cost of this facility would be more than $\$10^{12}$, and about 1 kg per year of antimatter could be produced. Also in 1982, R. R. Zito applied cryogenic confinement to design a demonstration matter/antimatter reactor for space-propulsion applications.

In his 1982 contribution, Forward examined aspects of the exhaust from a matter/antimatter rocket. The first particles to appear after protons and antiprotons annihilate are pions. From the point of view of an unaccelerated observer, these electrically-charged particles travel an average of 21 m before they decay into muons. Because of their high velocity, the muons travel about 2 km before they decay into electrons, positrons and neutrinos. Farther downstream from the spacecraft reaction chamber, the electrons and positrons interact to produce annihilation gamma rays.

Brice Cassenti (1982) investigated the efficiency of an antimatter rocket if it focuses pions or muons by magnetic nozzles. If pions are focused, as much as 67% of the energy released in the proton–antiproton interaction can be transferred to exhaust kinetic energy. If muons are focused, this efficiency is about 40%. Working with these efficiency factors, D. L. Morgan estimated that the exhaust velocity of a pion-relecting matter/antimatter rocket could be in excess of 0.9 c, and the vehicle acceleration could approximate 0.01 g.

Cassenti also investigated the kinematics of antimatter rockets. For velocity increments less than about 0.5 c, the ratio of reaction mass to fuel mass is about 4, since the ratio of antimatter to matter mass can be optimised. Forward (1982) applied this to determine that for velocity increments less than about 0.3 c, the optimum antimatter mass is always less than 1% of the total spacecraft launch mass. Forward (1982) also presented a simple formula relating the antimatter fuel mass ($M_{f,am}$) to unfuelled ship mass (M_0):

$$\frac{M_{f,am}}{M_0} = 0.9 \left(\frac{V_{\text{fin}}}{c} - \frac{V_{\text{in}}}{c} \right)^2 \tag{6.12}$$

where V_{in} and V_{fin} are respectively the spacecraft velocities at the beginning and end of antimatter-rocket operation.

Exercise 6.6 An antimatter rocket with an unfuelled mass of 10^7 kg is to be accelerated from rest to 0.1 c. The total fuel mass (from Cassenti (1982)) will be about 4×10^7 kg. Apply equation (6.12) to determine how much antimatter is required; then use an optimistic estimate of antimatter costs from *The Starflight Handbook*, $\$10^{10}$ per gm, to estimate the mission cost.

As part of the NASA-supported Advanced Space Propulsion effort, recent researchers have attempted to reduce the costs of antimatter propulsion so that serious mission planning can begin. Schmidt *et al.* have recently examined the cost savings if tiny amounts of antimatter are used to initiate fission and fusion reactions in much more massive micropellets. Such a strategy could ultimately reduce the antimatter fuel cost of an antimatter-propelled interstellar precursor mission to about $60 million. The specific impulse for such an antimatter interstellar-precursor mission is in the range 13,500–67,000 s.

As discussed by Schmidt *et al.* in 1999, the current world production rate for antimatter is 1–10 nanograms per year. Billions of dollars of investment would be required to obtain milligrams per year and reduce antimatter cost to trillions of US dollars per gramme. Because of its extreme volatility, this hazardous substance might be 'mass' produced in orbital or lunar antimatter factories.

Schmidt *et al.* also reviewed recent research on antimatter-rocket designs, which could be used if the cost of antimatter drops dramatically. One of these – the beam-core engine (Figure 6.5) – could have a specific impulse as high as 10^7 s, with 60% of the reaction energy transferred to the pion exhaust. The vehicle structure for a beam-core engine would be about 20% of the propellant mass.

As well as engine design and antimatter production, recent researchers have considered approaches to long-term antimatter storage. In a 1999 presentation by G. Gaidos *et al.*, the discussion included a portable penning trap (Figure 6.6) that can store up to 330 million ions per cubic centimetre in a 3.5-kiloGauss magnetic field. As discussed by S. D. Howe and G. A. Smith, higher-density antimatter storage for true interstellar missions might apply storage-ring techniques being developed for advanced particle accelerators.

Antimatter might remain too expensive a resource for human-occupied missions to the stars, but could antimatter ships be miniaturised for application with microtechnology or nanotechnology? This question has recently been investigated in papers published by R. A. Lewis *et al.* and G. Gaidos *et al.* Fermilab, in Batavia,

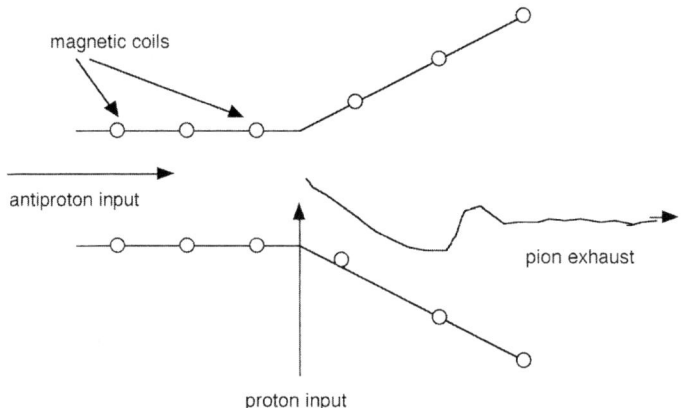

Fig. 6.5. The beam-core engine: one type of matter/antimatter annihilation rocket.

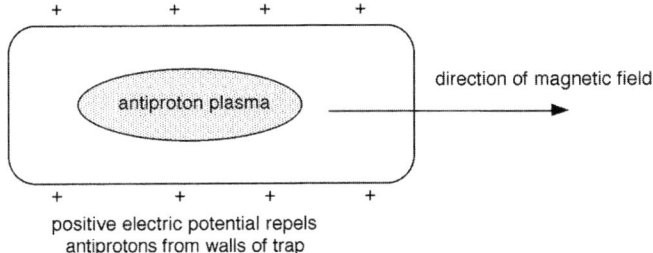

Fig. 6.6. A penning trap utilises electric and magnetic fields to contain antiprotons.

Illinois, produces 5–10 nanogrammes of antimatter per year, and in the near future this production rate could be raised by a factor of about 10, at a cost of about 10^8 antiprotons per dollar. A scheme involving antiproton-induced fission/fusion could conceiveably boost a small spaceprobe to a velocity of 129 km s^{-1}. In this approach, a small antiproton burst is directed at a small pellet of fissile material. The several-thousand electron volts released in this antimatter-induced fission could in turn be used to ignite a larger fusion micropellet.

In a follow-up paper, G. Gaidos *et al.* described how spacecraft mass could be reduced to about 400 kg, and a velocity increment of about 1,000 km s^{-1} could be achieved with a very small antimatter requirement. Such a craft could traverse 10,000 AU in 50 years. Further efficiencies could be obtained if the fusion fuel cycle were aneutronic.

In a 1999 paper, R. J. Halyard considered antimatter-assisted missions to nearby stars that could be accomplished using projections of current technology. An α Centauri flyby mission might require about a century for a 10,000-kg payload. A multistage mission to orbit Barnard's Star (about 6 light years distant) with a 10^5-kg payload would require an interstellar cruise of almost three centuries. Increased efficiency and lower costs for antimatter production are essential if these travel times are to be reduced.

As noted in *Science* (**284**, pp. 1597–1598, (1999)), NASA administrator Dan Goldin has proposed a research partnership between high-energy physicists and NASA. One of the fruits of this initiative might be greatly increased efficiency and reduced cost in the mass production of antimatter. Such a development will hasten the development of a true interstellar capability.

6.6 BIBLIOGRAPHY

Aston, G., 'Electric Propulsion: A Far-Reaching Technology', *Journal of the British Interplanetary Society*, **39**, 503–507 (1986) (also published as AIAA 85-2028).

Bond, A., Martin, A. R., Buckland, R. A., Grant, T. J., Lawton, A. T., Mattison, H. R., Parfatt, J. A., Parkinson, R. C., Richards, G. R., Strong, J. G., Webb, G. M., White, A. G. A. and Wright, P. P., 'Project Daedalus: The Final Report on the BIS Starship Study', supplement to *Journal of the British Interplanetary Society*, **31**, S1–S192 (1978).

Bond, R. A. and Martin, A. R., 'Deep Space Missions Using Advanced Ion Thruster and Nuclear Power Sources', IAA-92-0229.

Bussard, R. W., 'Some Physics Considerations of Magnetic Inertial-Electrostatic Confinement: a New Concept for Spherical Converging-Flow Fusion', *Fusion Technology*, **19**, 273–293 (1991).

Cassenti, B. N., 'Design Considerations for Relativistic Interstellar Rockets', *Journal of the British Interplanetary Society*, **35**, 396–404 (1982).

Chapline, G., 'Antimatter Breeders?', *Journal of the British Interplanetary Society*, **35**, 423–424 (1982).

Dyson, F., 'Interstellar Transport', *Physics Today*, **21**, No.10, 41–45 (October 1968).

Fearn, D., 'The Ultimate Performance of Gridded Ion Thrusters for Interstellar Missions', presented at STAIF 2000 Conference, University of New Mexico, Albuquerque, NM, January 30–February 3, 2000.

Forward, R. L., 'Antimatter Propulsion', *Journal of the British Interplanetary Society*, **35**, 391–395 (1982).

Forward, R. L. and Davis, J., *Mirror Matter*, Wiley, New York (1988).

Gaidos, G., Lawlor, L., Lewis, R. A., Scheidemantel, T. J. and Smith, G. A., 'The JPL/Penn State Portable Antiproton Trap', presented at NASA/JPL/MSFC/AIAA Annual Tenth Advanced Space Propulsion Workshop, Huntsville, AL, April 5–8, 1999.

Gaidos, G., Lewis, R. A., Meyer, K., Schmidt, T. and Smith, G. A., 'AIMStar: Antimatter Initiated Microfusion for Precursor Interstellar Missions', in *Missions to the Outer Solar System and Beyond, 2nd IAA Symposium on Realistic Near-Term Scientific Space Missions*, ed. G. Genta, Levrotto & Bella, Turin, Italy (1998), pp. 111–114.

Halyard, R. J., 'Antimatter Assisted Inertial Confinement Fusion Propulsion for Interstellar Missions', *Journal of the British Interplanetary Society*, **52**, 429–433 (1999).

Howe, S. D. and Smith, G. A., 'Development of High-Density Antimatter Storage', presented at NASA/JPL/MSFC/AIAA Annual Tenth Advanced Space Propulsion Workshop, Huntsville, AL, April 5–8, 1999.

Hyde, R., Wood, L. and Nuckolls, J., 'Prospects for Rocket Propulsion with Laser Induced Fusion Microexplosions', AIAA paper No. 72-1063 (December 1972).

Kammash, T. and Lee, M. J., 'A Near-Term Fusion Propulsion System for Interstellar Space Exploration', in *Missions to the Outer Solar System and Beyond, 1st IAA Symposium on Realistic Near-Term Scientific Space Missions*, ed. G. Genta, Levrotto & Bella, Turin, Italy (1996), pp. 263–271.

Kammash, T. and Lee, M. J., 'A Fusion Propulsion System for Near-Term Space Exploration', *Journal of the British Interplanetary Society*, **49**, 351–356 (1996).

Lewis, R. A., Smith, G. A., Cardiff, E., Dundore, B., Fulmer, J., Watson, B. J. and Chakrabarti, S., 'Antiproton-Catalyzed Microfission/Fusion Propulsion Systems for Exploration of the Outer Solar System and Beyond', in *Missions to the Outer Solar System and Beyond, 1st IAA Symposium on Realistic Near-Term Scientific Space Missions*, ed. G. Genta, Levrotto & Bella, Turin, Italy (1996), pp. 251–262.

Lipinski, R. J., Lenard, R. X., Wright, S. A. and West, J. L., 'Fission-Powered Interstellar Precursor Missions', presented at NASA/JPL/MSFC/AIAA Annual Tenth Advanced Space Propulsion Workshop, Huntsville, AL, April 5–8, 1999.

Massier, P. F., 'The Need for Expanded Exploration of Matter–Antimatter Annihilation for Propulsion Applications', *Journal of the British Interplanetary Society*, **35**, 387–390 (1982).

Matloff, G. L. and Chiu, H. H., 'Some Aspects of Thermonuclear Propulsion', *J. Astronautical Sciences*, **18**, 57–62 (1970).

Miley, G. H., Gu, Y., DeMora, J. M., Stubbers, R. A., Hochberg, T. A., Nadler, J. H. and Anderl, R. A., 'Discharge Characteristics of the Spherical Inertial Electrostati Confinement (IEC) device', *IEE Transactions on Plasma Science*, **25**, 733–739 (1997).

Miley, G. H., Nadler, J., Jurczyk, B., Stubbers, R., DeMora, J., Chacon, L. and Nieto, M., 'Issues for Development of Inertial Electrostatic Confinement (IEC) for Future Fusion Propulsion', AIAA-99-2140.

Morgan, D. L., 'Concepts for the Design of an Antimatter Annihilation Rocket', *Journal of the British Interplanetary Society*, **35**, 405–412 (1982).

Noble, R. J., 'Radioisotope Electric Propulsion of Sciencecraft to the Outer Solar System and Near-Interstellar Space', in *Missions to the Outer Solar System and Beyond, 2nd IAA Symposium on Realistic Near-Term Scientific Space Missions*, ed. G. Genta, Levrotto & Bella, Turin, Italy (1998), pp. 121–125.

Rostoker, N., Binderbauer, M. W. and Monkhorst, H. J., 'Colliding Beam Fusion Reactor', *Science*, **278**, 1419–1422 (1997).

Sanger, E., *Spaceflight: Countdown for the Future*, McGraw-Hill, New York (1964).

Schmidt, G. R., Gerrish Jr., H. P., Martin, J. J., Smith, G. A. and Meyer, K. J., 'Antimatter Production for Near-Term Propulsion Applications', presented at NASA/JPL/MSFC/AIAA Annual Tenth Advanced Space Propulsion Workshop, Huntsville, AL, April 5–8, 1999.

Shepherd, L. R., 'Performance Criteria of Nuclear Space Propulsion Systems', *Journal of the British Interplanetary Society*, **52**, 328–335 (1999).

Shmatov, M. I., 'Space Propulsion Systems Utilizing Ignition of Microexplosions by Distant Microexplosions and Some Problems Related to Ignition of Microexplosions by Microexplosions', *Journal of the British Interplanetary Society*, **53**, 62–72 (2000).

Solem, J. C., 'Medusa: Nuclear Explosive Propulsion for Interplanetary Travel', *Journal of the British Interplanetary Society*, **46**, 21–26 (1993).

Vulpetti, G., 'Antimatter Propulsion for Space Exploration', *Journal of the British Interplanetary Society*, **39**, 391–409 (1986).

Winterberg, F., 'Rocket Propulsion by Thermonuclear Microbombs Ignited with Intense Relativistic Electron Beams', *Raumfahrtforschung*, **15**, 208–217 (1971).

Winterberg, F., 'Rocket Propulsion by Staged Thermonuclear Microexplosions', *Journal of the British Interplanetary Society*, **30**, 333–340 (1977).

Zito, R. R., 'The Cryogenic Confinement of Antiprotons for Space Propulsion Systems', *Journal of the British Interplanetary Society*, **35**, 414–421 (1982).

7

Twenty-first century starflight

> *Looking out from the continent-sized cities and vast game preserves that may be our future on this planet, youngsters will dream that when they are grown, if they are very lucky, they will catch the night freight to the stars.*
>
> Carl Sagan, *The Cosmic Connection* (1973)

By the end of the twenty-first century we will have imaged Earth-like planets (if these worlds exist) orbiting nearby stars, and our probes will have reached the heliosphere at, at least, the inner fringes of the Oort Cloud. Fission propulsion will probably be incapable of carrying humans to these beckoning new worlds; fusion-pulse may remain politically and socially unacceptable, solar-sailing is too slow, and antimatter is too expensive. This will be a time of vast change on Earth as our population peaks and nation states begin to give way to a true global civilization. Is there any hope, then, for a propulsion system that could take at least a few humans to habitable worlds orbiting nearby stars, on missions that begin late in the twenty-first century?

As it happens, there is one star-travel approach that may become feasible by the middle of the century. Figure 7.1 presents the various aspects of this approach – beamed-energy sailing. A solar-powered station is located closer to the Sun than the starship. Solar collectors focus sunlight on a solar-pumped beam-projector attached to the solar-collector array. A laser beam or particle beam is generated by the power station. This is directed at the distant starship with the beam divergence angle shown. The starship, which consists of payload attached to a sailcraft, accelerates to its cruise velocity by the exchange of momentum with the impinging beam.

As the starship approaches the destination star, it applies a magnetic field (or magsail) for the first stage of deceleration. The magsail reflects interstellar ions, acting as a magnetic dragbrake. For the final deceleration stage, the craft decelerates to planetary velocities using a solar sail directed at the destination star. Similar magnetic techniques may have application in changing spacecraft direction in interstellar space without application of thrust and even to supplying onboard power during the interstellar cruise phase.

Acceleration

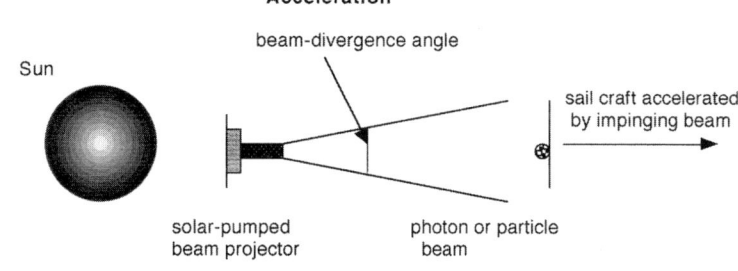

Deceleration

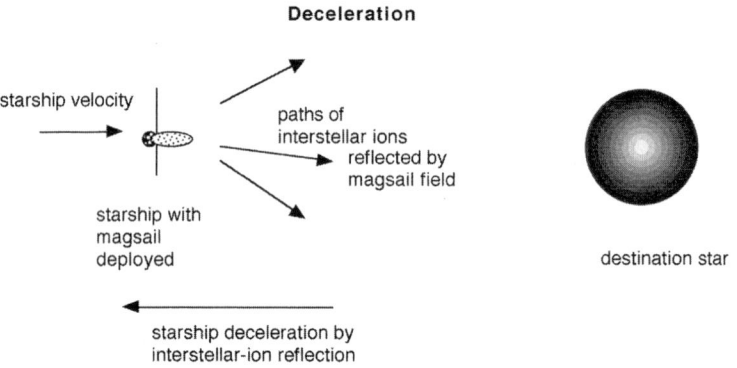

Fig. 7.1. Aspects of beamed-energy sailing.

For acceleration, laser-beam photons carry less momentum than do accelerated particles, and hence are less efficient. Because accelerated particles generally are electrically charged, the beam divergence of particle beams will be greater than that of laser beams. Techniques have been suggested to reduce the beam divergence of both laser and particle beams.

7.1 LASER/MASER SAILING FUNDAMENTALS

Perhaps because particle-beam techniques are of military significance in missile defence schemes, more research has been published on optical laser or microwave laser (maser) application to spacecraft acceleration. This work leans heavily upon studies of solar power stations (SPS) in space that beam energy to Earth by microwave beams. Even an early SPS has the potential to accelerate a nanominiaturised probe to respectable interstellar velocities.

We might instead consider a group of simple reflectors orbiting the Sun to focus sunlight directly on a distant sail-equipped starship and thereby avoid the difficulties inherent in designing a high-tech laser/maser or particle-beam projector. But the Sun is a source with finite extent rather than being a point object. As discussed by Matloff in 1996, interstellar application of such light-concentrators will therefore probably be very limited.

Instead, we consider a solar array with a radius of R_{array} directing its light into a laser (or maser). The collected sunlight might first be converted into electricity, or alternatively it might be used directly to pump the laser. The efficiency of sunlight conversion to a collimated EM radiation beam is ε_{las}. If the separation between the Sun and the solar-pumped laser power station is $R_{sl,au}$ in Astronomical Units, we can estimate laser power using the 1996 formalism of Matloff and Potter:

$$P_{laser} = \frac{1400}{R_{sl,au}^2} \varepsilon_{las} \pi R_{array}^2 \text{ W} \quad (7.1)$$

where R_{array} is the (disc-shaped) collecting array's radius in metres, and 1,400 W m^{-2} is the solar irradiance on an object 1 AU from the Sun, oriented so that the sunlight is normal to the object. The laser power station may be light-levitated in a stationary position between the Sun and the starship, orbiting the Sun or in a parabolic solar orbit following the starship at a slower speed.

If all the laser light is incident on the fully opaque starship sail, the acceleration of the interstellar spacecraft depends upon sail reflectivity to the laser beam, REF_{sail}, ship mass in kilogrammes, M_s, laser-beam power, and the speed of light c in metres per second, in a manner analogous to the acceleration of the solar sail, as considered in Chapter 4:

$$ACC_{laser-sail} = \frac{(1 + REF_{sail})}{M_s c} P_{laser} \text{ m s}^{-2} \quad (7.2)$$

In the case of partially transmissive sails (as considered later in this chapter), the factor $(1 + REF_{sail})$ in equation (7.2) will be replaced by $ABS_{sail} + 2REF_{sail}$. Where ABS_{sail} is sail fractional absorption, as discussed in Matloff (1995).

As discussed in many references, including Matloff and Potter (1996), the laser beam's divergence angle is governed by Rayleigh's criterion. We can relate laser wavelength λ_{laser} in metres to the diameter of the laser-transmitting optics in metres $D_{las-tran}$, and the separation in metres between the laser power station and the starship $D_{las-ship,max}$ at which the laser beam completely fills the (disc-shaped) sail with radius R_{sail}:

$$\frac{2.44 \lambda_{laser}}{D_{las-tran}} = \frac{2 R_{sail}}{D_{las-ship,max}} \quad (7.3)$$

Starship acceleration can continue for laser-ship separations greater than $D_{las-ship,max}$, albeit at a reduced rate. For greater separations, starship acceleration will vary with the inverse square of the distance to the power station.

Let us examine how we can design an interstellar light-sailing probe using equations (7.1), (7.2) and (7.3). First assume that the laser power station is 'parked' between the Sun and the probe in a stationary position (from the starprobe's point of view) 0.1 AU from the Sun's centre. If the laser efficiency is 0.3 and the solar-array collecting radius is 100 m, solution of equation (7.1) reveals that the power in the laser beam is about 1.3×10^9 W.

If the sail reflectivity is 0.9 and the total probe mass is 50 kg, substitution of these values and the above value for beam power in equation (7.2) yields an acceleration of 0.16 m s^{-2}, or about 0.017 g. Starting from rest and accelerating at this rate for two years, the probe reaches a terminal velocity of about 10^7 s^{-1}, or 0.035 c. This probe will reach α Centauri in about 125 years. Assuming constant acceleration, the probe's average velocity during the two-year acceleration period will be about 0.0175 c. The probe's distance from the Solar System ($D_{\text{las-ship,max}}$) will be about 0.035 light years, or 3.15×10^{14} m at the termination of laser-beam operation, assuming that the beam completely fills the sail at the end of laser-beam acceleration.

Assume next that the sail consists of 25-nm thick aluminium and has a mass of 27 kg. The area of this sail (calculated by dividing sail mass by the product of sail density and thickness) is $4 \times 10^5 \text{ m}^2$. Assuming a disc-sail configuration, the radius of the probe's sail, R_{sail}, is about 350 m.

If we next assume a yellow–green laser with a wavelength λ_{laser} of 0.5 μm (5×10^{-7} m) and substitute into equation (7.3), we find that in order to completely fill a 350-m radius sail at a distance of 3.15×10^{14} m with the laser beam, the diameter of the laser-beam transmitting optics ($D_{\text{las-tran}}$) is about 550 km.

Exercise 7.1 Photon beam dispersion is inferior at longer wavelengths, which affects interstellar communication as well as beam-propulsion studies. Apply equation (7.3) to estimate how much larger the transmitting optics must be to completely fill the 350-m radius sail at a distance of 0.035 light years from the Sun, if 1-μm or 10-μm infrared light is transmitted by the laser.

The 1.3×10^9 W beam power required to deliver a 27-kg payload to α Centauri in about 125 years approximates the power output of a large terrestrial electrical power station. Application of photon-power beaming to the acceleration of true starships on one-way interstellar journies will equal or exceed the current terrestrial electrical-power generating capacity (about 10^{13} W).

Unlike slower solar-sail starships that complete their acceleration in a relatively short time, laser sails must be exposed to occasional impacts by interstellar dust for decades. Geoffrey Landis (1999) has reported that dust erosion of interstellar light-sails may be less than previously expected.

Exercise 7.2 In 1975, Edward Gilfillan Jr estimated the non-fuel mass of a very minimal interstellar 'generation ship' or ark (crewed by about twenty people) at about 65,000 kg. If the sail and associated structure bring the interstellar ark mass up to 100,000 kg, estimate, from the above example, the laser-beam power required to deliver this craft to α Centauri in 125 years. How does this power requirement compare with the current terrestrial electrical-power generation estimate? Using the laser efficiency and position used in the example, how large is the size of the solar collecting-array for the laser power station? If you estimate the radius of a 35,000-kg, 25-nm thick aluminium sail from the above example, you will see that the starship's larger sail results in a reduced size for the laser transmitting optics.

But there is one major problem with the application of power-beam technology to true interstellar flight. To paraphrase Robert Zubrin in his 1999 book, decelerating using power-beams alone is very, very difficult in principle, and perhaps impossible in practice.

7.2 STARSHIP DECELERATION USING THE MAGSAIL

A potential method of starship deceleration that does not strain the limits of contemporary technology does, however, exist. The principle of this device – a form of magnetic braking called the 'magsail', – is illustrated in Figure 7.2.

The magsail is derived from early approaches to the collection of fuel for interstellar ramjets, which are described in a following chapter. In 1974, Gregory Matloff and Alphonsis Fennelly suggested that the magnetic field generated by a superconducting solenoid could extend for thousands of kilometres in the interstellar medium. Gyrating around the ship-generated magnetic field lines, interstellar ions would approach the ship closely. But Matloff and Fennelly were unable to conclude whether the ions would be collected into the scoop or reflected back into space.

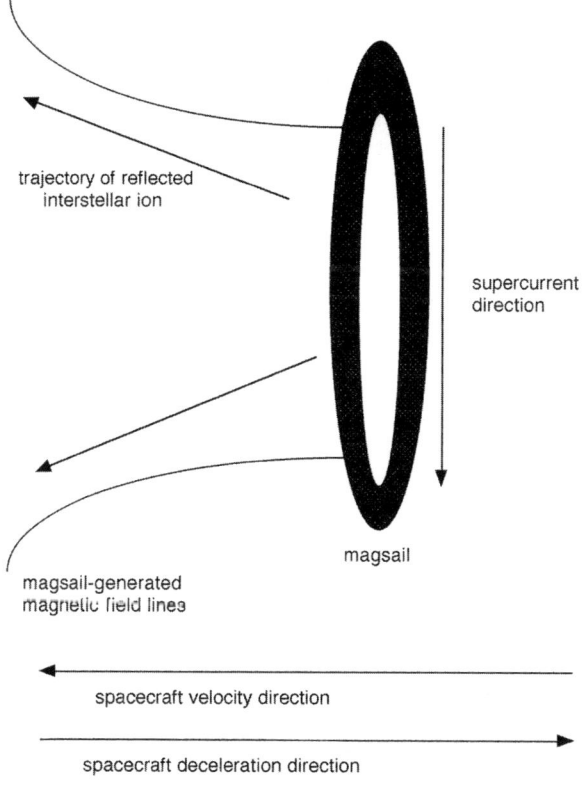

Fig. 7.2. The magsail as an interstellar braking mechanism.

In 1990 this concept was reinvestigated by Dana Andrews and Robert Zubrin. Using sophisticated computer codes, these authors demonstrated that a solenoidal field would tend to reflect interstellar ions elastically, decelerating the ship by linear momentum conservation. Their analysis also reveals that a simple supercurrent ring is superior as a braking mechanism to a superconducting solenoid.

The magsail principle works as follows. Magnetic field lines emerge from the supercurrent ring, as shown in Figure 7.2. Interstellar ions gyrate around the field lines, and therefore approach the starship. Close to the starship, the magsail-generated magnetic field lines are so close together that a magnetic-mirror effect is created. The ions are reflected back into space, in the direction of starship velocity, and the spacecraft decelerates.

In interplanetary space the effective field radius of the magsail is about ten times its physical radius. The magsail's mass can be estimated using equation (9) of the Zubrin/Andrews 1989 AIAA paper as:

$$M_{mag} = 2\pi R_{mag} \frac{I_{mag}}{J_{pm,mag}} \quad (7.4)$$

where R_{mag} is the magsail's physical radius, I_{mag} is the magsail supercurrent, and $J_{pm,mag}$, is the magsail's current/mass density ration in amp-m kg.

To obtain an expression for magsail-induced deceleration DEC_{mag}, we next modify equation (10) of Zubrin/Andrews (1989) to include the ratio of magsail mass to ship mass M_s:

$$DEC_{mag} = -0.59 \left(\mu_{fs} \rho_{in}^2 V_s^4 \frac{R_{mag}}{I_{mag}} \right)^{0.33} J_{pm,mag} \frac{M_{mag}}{M_s} \quad (7.5)$$

where μ_{fs} is the permeability of free space ($4\pi \times 10^{-7}$ N amp^2), ρ_{in} is the interstellar-ion mass density and V_s is the ship velocity. Integration of equation (7.5) yields a non-relativistic expression for ship velocity V_{mag} at time t during magsail deceleration as a function of ship velocity during interstellar cruise V_{cr}, at the start of magsail deceleration:

$$V_{mag} = \left[\frac{1}{0.197 \left(\mu_{fs} \rho_{in}^2 \frac{R_{mag}}{I_{mag}} \right)^{0.33} J_{pm,mag} \frac{M_{mag} t}{M_s} + \frac{1}{V_{cr}^{0.33}}} \right]^3 \quad (7.6)$$

From Zubrin/Andrews (1989), magsail mass $= 8.7 \times 10^4$ kg, and magsail supercurrent $= 159$ kiloamperes. If we adopt a current/density mass ratio of 5×10^6 amp-m kg^{-1}, a compromise between Zubrin/Andrews near-term and optimistic projections, equations (7.4) yields a magsail physical radius of about 420 km.

In operation, the effective magsail physical radius will be reduced because of the necessity to store supercurrent to compensate for variations in interstellar ion density. Since this might be done by counter-winding some coils to the main magsail, we assume that R_{mag} is reduced to 120 km. Selecting a conservative value of local interstellar-medium ion density (0.05 protons cm^3, from Matloff/Fennelly (1974)),

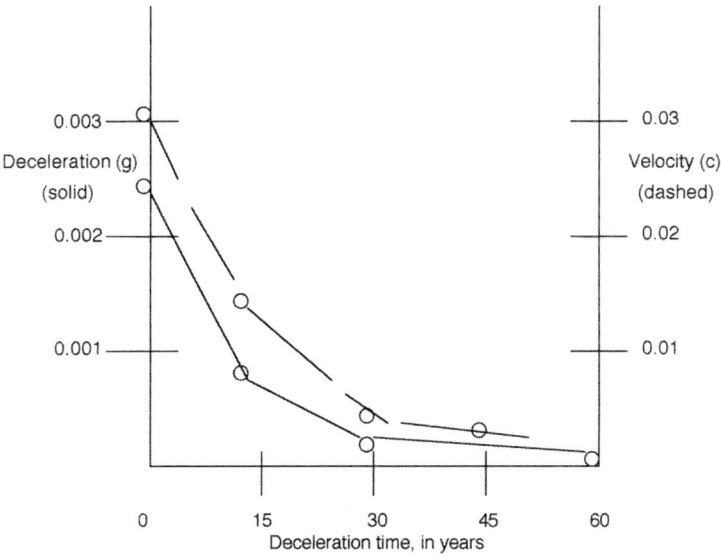

Fig. 7.3. A typical magsail deceleration profile. Deceleration is denoted as fractions of Earth's surface gravity; velocity is denoted as fractions of the speed of light.

and assuming a total ship mass of 3.8×10^5 kg, we can solve equations (7.5) and (7.6).

Figure 7.3 presents a magsail deceleration profile for the parameters described above, and an interstellar cruise velocity of 0.03 c, obtained by the solution of equations (7.5) and (7.6). The ship requires about 50 years to decelerate from 0.03 c to 0.0022 c, during which time it traverses about 0.46 light years. Deceleration from 0.03 c to 0.0016 c requires about 60 years, during which time the ship traverses about 0.48 light years. Deceleration from 0.04 c adds just a few years to the deceleration process.

Note that magsail deceleration is much more efficient at high velocities. For this reason the terminal stage of starship deceleration will use the light sail directed at the target star, as described in a previous chapter. The above analysis may, however, actually underestimate magsail performance, since a magsail could be pointed into the solar wind of the destination star, thereby enhancing its performance. Also, neutral interstellar atoms encountering the rapidly varying magsail magnetic field might be ionised.

As reviewed in 1997 by Cocks *et al.*, many inner Solar System applications of superconducting magsails have been suggested. But as discussed by Vulpetti and Pecchioli, these may be difficult to implement, because present-day superconductors operated within the orbit of Mars tend to 'go normal' and lose superconductivity unless massive thermal shielding is employed.

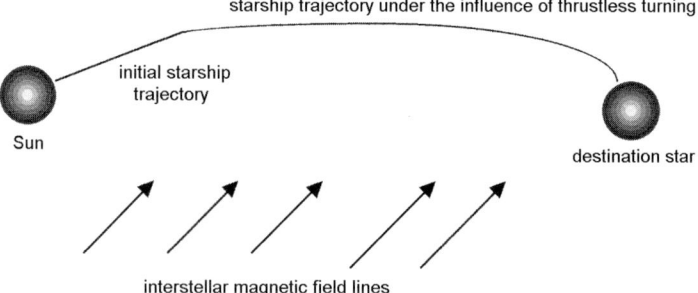

Fig. 7.4. The application of thrustless turning to alter a starship's trajectory.

7.3 THRUSTLESS TURNING

Some of the earliest papers dealing with photon-beamed space travel speculated some form of thrustless manoeuvres using the interstellar magnetic field. These would allow a ship to circle back and re-enter a poorly collimated beam, or even serve as an abort mode to return a failed interstellar mission to the Solar System without great fuel expenditure.

Figure 7.4 presents the application of interstellar thrustless turning. A starship travels through the interstellar magnetic field. An electrostatic charge or magnetic field on the spacecraft will alter the craft's trajectory without the expulsion of propellant. In a constant interstellar magnetic field, the spacecraft will follow a circular trajectory.

The earliest approaches to thrustless turning in the literature are Forward (1964) and Norem (1969), both of which considered Lorentz-force turning of an electrostatically charged spacecraft. In this approach, a starship with mass M_s cruises through a constant interstellar magnetic field with intensity B_{ism}. The spacecraft's velocity relative to the magnetic field lines is V_{srm}, and a net electrical charge Q_{net} is carried by the spacecraft (possibly generated by the decay of radioactive isotopes). Since the Lorentz force and centripetal force will be equal,

$$Q_{net} V_{srm} \times B_{ism} \cong Q_{net} B_{ism} V_{srm,per} \cong \frac{M_s}{R_{estr}} V_{srm}^2 \qquad (7.7)$$

where $V_{srm,per}$ is the velocity component of the spacecraft perpendicular to the local interstellar magnetic field, R_{estr} is the electrostatic turning radius of the spacecraft, and the interstellar magnetic field is considered to be approximately perpendicular to the spacecraft's velocity vector. Rearranging equation (7.7), we find

$$R_{estr} \cong \frac{M_s V_{srm}^2}{Q_{net} B_{ism}} \qquad (7.8)$$

Exercise 7.3 A starship with a mass of 5×10^6 kg is cruising through the interstellar medium with a velocity of 0.02 c. From Pikel'Ner (1968), the magnitude of the interstellar magnetic field in the Galactic vicinity of the Sun is

Sec. 7.3]   **Thrustless turning**   91

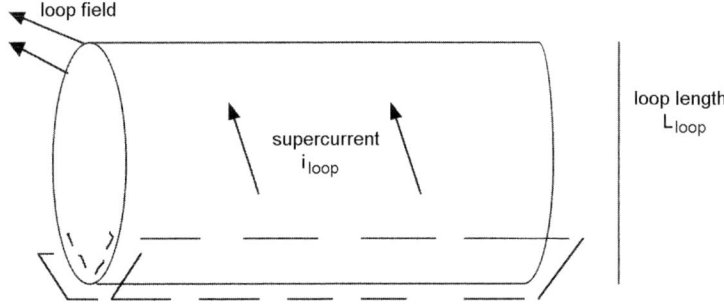

Fig. 7.5. One approach to thrustless electrodynamic turning requires a partially sheathed superconductor. The interstellar magnetic field is assumed to be perpendicular to supercurrent and loop length. Inner and outer superconducting shields are denoted by dashed lines.

about 5×10^{-10} webers m^2. If the ship carries a net electric charge of 5×10^6 c, what is the electrostatic turning radius from equation (7.8). To turn through an angle of 180°, the ship traverses half of the circumference of a circle with radius R_{estr}. At a ship velocity of 0.02 c, how much time is required to complete a 180-degree thrustless turn?

Ships of reasonable size will require enormous electrostatic charges to complete turning manoeuvres within decades. One obstacle to the utilisation of such enormous net charges is Debye–Hückel screening, described in Jackson (1962). Interstellar ions with charges opposite to the spacecraft will be attracted from great distances, thereby reducing the effective spacecraft charge. Application of Debye–Hückel theory to this situation is non-trivial because of the starship's high velocity and the tenuous ion density of the interstellar medium. Electrodynamic turning is suggested as a possible alternative to electrostatic turning.

An approach to electrodynamic thrustless turning was presented by Matloff *et al.* in 1991. As shown in Figure 7.5, a unidirectional supercurrent sheet is first created by partially shielding a superconducting loop or solenoid with an external layer of superconductor through which the interstellar magnetic field cannot penetrate. If the current in the superconducting loop or solenoid is i_{loop} and the length of the loop is L_{loop} (as shown in Figure 7.5), a magnetic force $F_{loop,mag}$ causes the loop to turn through a circular trajectory in the (constant) interstellar magnetic field:

$$F_{loop,mag} = i_{loop} L_{loop} \times B_{ism} \cong i_{loop} L_{loop} B_{ism} \cong \frac{M_s}{R_{mtr}} V_{srm}^2 \qquad (7.9)$$

where L_{loop} is approximately perpendicular to the interstellar magnetic field and R_{mtr} is the radius of magnetic turn. Rearrangement of equation (7.9) yields the following result for magnetic turn radius:

$$R_{mtr} \cong \frac{M_s V_{srm}^2}{i_{loop} L_{loop} B_{ism}} \qquad (7.10)$$

In an effort to estimate reasonable values for the parameter $i_{loop} L_{loop}$ in equation (7.9), Matloff et al. (1991) applied the analysis to a thin-film superconducting solenoid such as that investigated by Matloff and Fennelly (1974) in the case of a solar-sail starship designed to utilise thrustless turning to make multiple passes of the Sun. They utilised standard calculations of membrane stress, and assumed that refurled cable and sail could be utilised to compensate for magnetic hoop stress during a thrustless turn. For a payload mass of 5×10^6 kg and a total ship mass of 3.5×10^7 kg, the value calculated for the current-length kg of about 1.5×10^6 amp-metre kg^{-1}. This resulted in a value for the product $i_{loop} L_{loop}$ of about 1.5×10^{14} amp-metre. Utilisation of the more conservative bulk-superconductor approximation of Zubrin/Andrews (1989) resulted in a value of $i_{loop} L_{loop}$ about 50% lower than the value mentioned above.

Exercise 7.4 For the spacecraft mass $i_{loop} L_{loop}$ value cited above and the interstellar magnetic field strength used in Exercise 7.3, calculate turning radius and time for a 180-degree electrodynamic thrustless turn by a starship moving at 0.005 c.

7.4 PERFORATED LIGHT SAIL OPTICAL THEORY

One method of increasing a light-sail's performance is to use thrustleness turning to make multiple passes through a laser beam or by the Sun. Another approach is to reduce sail mass as much as possible. Basing his work upon unpublished results of Freeman Dyson, in 1985 Robert Forward suggested one possible method of greatly reducing light-sail mass.

This method, called the 'perforated light sail', is illustrated in Figure 7.6, which shows a mesh constructed of rectangular conducting wires. Mesh parameters include mesh wire thickness t_{mesh}, mesh wire width $2a_{mesh}$, and parameter g_{mesh}, which equals $2a_{mesh}$ + wire separation.

In 1995, Matloff presented a theory of mesh optical properties based upon an approach in Driscoll and Vaughan (1978). This theory applies if rectangular wire cross-sectional circumference, u_{mesh}, is defined as $8a_{mesh}$ and if the mesh satisfies the following conditions:

light wavelength $\lambda \gg 2g_{mesh} > 16a_{mesh}$
$a_{mesh} >$ skin depth (δ_{skin})
$t_{mesh}/g_{mesh} \ll 1$

For meshes that satisfy these conditions, mesh spectral fractional transmission ($T_{\lambda,mesh}$), spectral fraction absorption ($A_{\lambda,mesh}$), and spectral fractional reflectance ($R_{\lambda,mesh}$) at wavelength λ can be related:

$$T_{\lambda,mesh} \cong \frac{4g_{mesh}^2}{\lambda^2} \left\{ Ln \left[\sin \left(\frac{\pi a_{mesh}}{g_{mesh}} \right) \right] \right\}^2$$

$$A_{\lambda,mesh} \cong \frac{2g_{mesh} R_{\lambda,mesh}}{u_{mesh}} \left(\frac{c}{\sigma_{mesh} \lambda} \right)^{1/2}$$

(7.11)

Sec. 7.4]   Perforated light sail optical theory   93

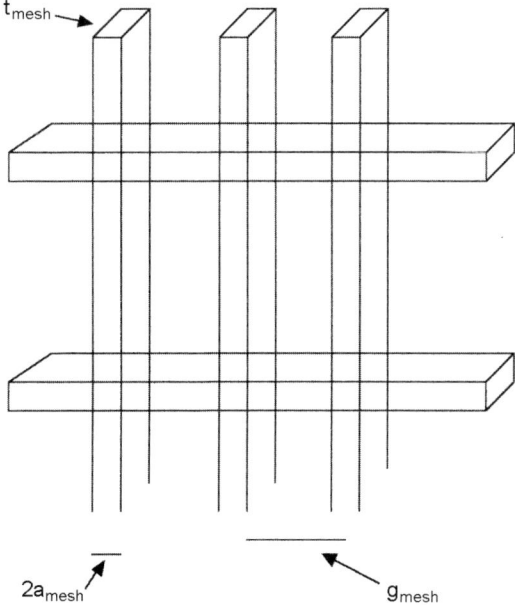

Fig. 7.6. A perforated solar sail constructed from rectangular wires.

where c is the speed of light, and σ_{mesh} is the mesh wire conductance in reciprocal seconds. For the theory to apply, $1 - T_{\lambda,mesh} \approx R_{\lambda,mesh}$ since $R_{\lambda,mesh} \gg A_{\lambda,mesh}$.

As discussed in Jackson (1962), for a good conductor such as room-temperature nickel, $\sigma_{mesh} \approx 10^{-17}\,\mathrm{s}^{-1}$. Also from Jackson, the skin depth (δ_{skin}) is proportional to the inverse of the square root of the frequency of the electromagnetic radiation incident on the mesh. Jackson states that for frequencies of 60 Hz and 108 Hz respectively, the skin depth is about 0.85 cm and 0.71×10^{-4} cm for copper. For visible light with a frequency of about 6×10^{14} Hz, skin depth is about 3 nm.

Matloff (1995) compared predictions with measurements for a mesh described by Renk and Genzel (1962) for 100–250 μm infrared light. Agreement between theory and experiment was fairly close with the theory underestimating reflectance at low wavelengths. Theoretical values for spectral fractional absorptance were generally two to three times greater than the corresponding experimental values.

Admittedly, the theory presented here for optical performance of a perforated light sail is approximate and restrictive. Until a better theory is published, the spectral transmission and reflection results presented in Table 7.1 for a sample perforated sail, from Matloff (1995) should be treated as very approximate.

However, from these results the mesh considered is more reflective for red light than for blue light. One can easily wavelength average these optical parameters with the Sun's relative black-body emittance curve to determine the optical performance of a perforated solar sail in sunlight.

When considering performance of a partially transmissive light sail, the reflectivity factor for an opaque light sail ((1 + Reflectivity)/2) should be replaced by the

Table 7.1. Fractional spectral reflectance, absorption and transmission of an aluminium perforated light sail with $g_{mesh} = 150$ nm, $a_{mesh} = 15$ nm, and $\sigma^{1/2} = 3.05 \times 10^{-8}\,\text{s}^{-0.5}$

λ	$B_{\lambda,mesh}$	$A_{\lambda,mesh}$	$T_{\lambda,mesh}$
0.4 μ	0.22	0.10	0.68
0.5	0.50	0.20	0.30
0.6	0.65	0.24	0.11
0.7	0.75	0.25	0.00

From Matloff (1995)

expression ((Absorptivity + 2 × Reflectivity)/2). From Wolfe (1965), the emissivity for a non-opaque or transmissive light sail (see the earlier discussion on solar sails for more information on emissivity) can be defined as

$$\varepsilon_{trans,sail} = \frac{(1 + T_{trans,sail})(1 - R_{trans,sail})}{1 - T_{trans,sail}R_{trans,sail}} \quad (7.12)$$

where the subscript 'trans, sail' attached to the symbols for emissivity, transmission and reflection denote a partially transmissive sail.

Perforated light sails do seem to result in significant performance improvements, when compared with their non-perforated counterparts. A great deal more theoretical and experimental work can therefore be profitably devoted to this subject. Nanotechnologies can be developed to construct these structures. As discussed by Forward in his 1985 paper, very low-mass interstellar probes are possible if we can construct perforated light sails out of material that can also serve as electronic devices.

7.5 THE FRESNEL LENS: A METHOD OF IMPROVING LASER-BEAM COLLIMATION

In 1984, Robert Forward suggested another conceptual technique to improve the performance of an interstellar laser sail. As shown in Figure 7.7, Forward would position a thin-film refractive optical element – a Fresnel lens – between the laser and the starship. If the position of these three optical elements (laser, lens and starship) can be accurately maintained for decades or centuries (no mean task!), the light-years distant starship could be presented with a very well collimated beam, at a distance measured in light years.

Forward proposed a 1-μm wavelength solar-pumped laser (perhaps at the orbit of Mercury), with laser collimation such that the 500-km radius Fresnel zone lens at 15 AU (about 2.25×10^9 km) from the laser is completely filled by the laser beam.

A space Fresnel zone lens would be constructed using concentric rings of thin film

Sec. 7.5] The Fresnel lens: a method of improving laser-beam collimation 95

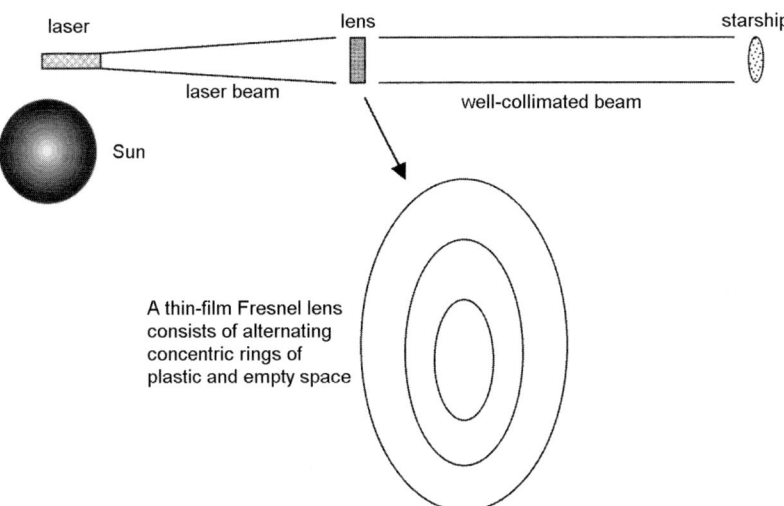

Fig. 7.7. A Fresnel lens in the outer Solar System can direct a laser beam to a light sail light years from the Sun.

plastic separated by rings of empty space. The number of zones in the Fresnel lens (NZ_{fl}) can be estimated using equation (18) of Forward (1984):

$$NZ_{fl} = \frac{R_{fl}^2}{f_{fl}\lambda} \qquad (7.13)$$

where R_{fl} is the radius of the Fresnel lens, f_{fl} is the focal length of the Fresnel lens, and λ is the wavelength of the laser light.

If the lens radius is 500 km and the lens focal length is 15 AU, about 110,000 Fresnel zones are required for 1-μm laser light. A well-collimated beam from the lens would completely fill the 500-km radius sail of a very distant starship if the laser were to be positioned 15 AU from the lens, as shown.

From equation (19) of Forward (1984), the radius of the first Fresnel zone can be calculated:

$$R_{fl,1} = (f_{fl}\lambda)^{1/2} \qquad (7.14)$$

For the case discussed above, $R_{fl,1}$ is approximately 1.5 km. Applying equation (24) of Forward (1984), the spacing between the outer Fresnel zones (or the width of the plastic in the outer zone) can be written as

$$S_{fl,\text{out}} = \frac{f_{fl}}{2R_{fl}}\lambda \qquad (7.15)$$

For the case considered above, $S_{fl,\text{out}}$ is about 2.25 m. From Forward (1984), the mass of the thin-film plastic required for this Fresnel lens is about 5×10^8 kg.

Exercise 7.4 In 1985, Eric Jones proposed a maser-propelled light sail pushed by a well-collimated 3-m wavelength microwave beam. How many Fresnel lens zones would be required for this wavelength if the maser-lens separation (lens focal length) is 15 AU and the lens radius is 3,000 km?

In 1999, Korman *et al.* concluded that even contemporary thin-film and inflatable space technologies are quite capable of application to the manufacture of large Fresnel lenses in space.

7.6 ROUND-TRIP INTERSTELLAR VOYAGES USING BEAMED-LASER PROPULSION

Forward's 1985 paper dealt with 'Starwisp', a maser-accelerated nanoprobe in which smart circuitry comprises perforated sail elements, and Jones' 1985 paper considered a maser-accelerated interstellar colonisation craft. Starwisp is undecelerated, and Jones' craft would require a magsail or some other method for deceleration. But as Norem pointed out in 1969 and Forward discussed in 1984, there are at least two possible ways to conduct round-trip interstellar journies using laser beams and light sails.

Norem would first accelerate his light sail using a laser beam and direct the spacecraft past the target star. Then, he would utilise thrustless turning in the interstellar magnetic field so that the starship changes direction by 180°. To decelerate, the craft would manoeuvre once again into the beam, this time moving towards the Sun. By the time it once again entered the destination solar system, its speed relative to the target star would be zero.

To return his craft to Earth, Norem would first accelerate the craft once again into the beam. After interstellar cruise velocity is achieved, thrustless turning would again be employed to redirect the craft by 180° so that it is travelling towards the Earth. Approaching our Solar System, Norem's craft would again enter the beam, to be decelerated by the time it reaches Earth.

Forward would launch a multi-stage light-sail from the Solar System, and first accelerate it to interstellar cruise velocity by the laser beam. Approaching the destination star system, the sail would separate into two sections. The larger and leading sail would act as a mirror to reflect the laser beam onto the starship's smaller sail for deceleration. After completion of their mission, the starship crew would redirect their craft into the beam reflected from the larger mirror sail, which by now might be light years distant. This reflected beam would accelerate their craft back towards Earth. Approaching the Solar System, the starship's sail would re-enter the laser beam for deceleration.

Needless to say, both approaches are enormously demanding in terms of beam aim and collimation and starship manoeuving. Even if the technical problems can be solved, round trips to even the nearest stars will take centuries for reasonable beam powers and ship masses.

Other researchers have suggested that we could reduce beam-power requirements by using the laser beam's energy to heat a working fluid instead of the beam momentum. Non-relativistic laser-heated interstellar rockets were suggested by John Bloomer in 1966; a relativistic treatment of laser-heated interstellar rockets was published by Jackson and Whitmire (1978); and Jordin Kare (2000) has considered applying this approach to near-term interstellar precursor probes. (To consider the non-relativistic kinematics of a laser-heated interstellar-electric rocket, refer to Section 4.2, and replace the Sun by the laser as a power source.)

In 1977, Whitmire and Jackson also published a relativistic treatment of a laser-heated interstellar ramjet. A non-relativistic treatment of this craft's kinematics is presented in the next chapter. Since the particles exhausted from rockets and ramjets have much higher linear momentum than do photons, beam power would be lower than for laser sailing craft. But when considering round-trip missions, laser-beam collimation and aim would be just as challenging!

7.7 INTERSTELLAR PARTICLE-BEAM PROPULSION

Another variation on the beamed-propulsion theme is to replace the laser-beam apparatus with a charged-particle beam accelerated by solar energy, as proposed by Clifford Singer in 1979. As shown in Figure 7.8, spacecraft acceleration would be effected by momentum transfer from the particle beam to the spacecraft.

Belbruno and Matloff considered application of particle-beaming to thin-film interstellar probes in 1993, and Matloff further investigated the non-relativistic performance of such craft in his 1996 'Robosloth' paper.

A Robosloth-type probe was assumed to be a thin-film spacecraft initially unfurled close to the Sun and accelerated by solar sail to about 0.005 c, on a trajectory that would require about 900 years to reach α Centauri. The craft is then inserted into a particle beam (before or after the solar pass) that increases its interstellar

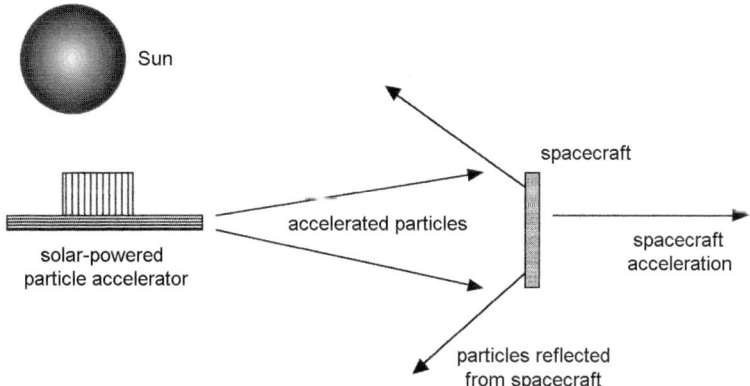

Fig. 7.8. Accelerating an interstellar spacecraft by momentum transfer from an accelerated particle beam.

cruise velocity. From Matloff's 1996 Robosloth paper, the change in ship velocity ΔV_s can be related to particle-beam velocity, V_{beam}:

$$\frac{\Delta V_s}{V_{beam}} = \frac{2\mu_{mte}\frac{M_{part}}{M_s}}{1 + 2\mu_{mte}\frac{M_{part}}{M_s}} \qquad (7.16)$$

where M_{part} is total particle-beam mass, M_s is the spacecraft mass, and μ_{mte} is the momentum transfer efficiency. For a fully elastic collision, $\mu_{mte} = 1$; $\mu_{mte} = 0.5$ for a fully inelastic collision. Since the beamed particles are electrically charged, a charged 'reflection plate' attached to the interstellar spacecraft should result in highly elastic collisions.

For $\mu_{mte} = 0.95$, $V_{beam} = 0.012\,c$, a total starprobe mass of 100 kg and a total beamed-particle mass of 73 kg striking the spacecraft, equation (7.16) can be solved to obtain a spacecraft velocity increment of $0.007\,c$. The total beam energy is about 5×10^{14} J.

Exercise 7.5 Estimate spacecraft velocity increment and beam energy for the same spacecraft and beam masses and beam velocity if the momentum-transfer efficiency is reduced to 0.7.

One problem with particle-beam momentum-transfer schemes is the fact that charged particle beams are much more poorly collimated than are laser or maser beams. One way to improve beam collimation, as suggested by Belbruno and Matloff, is to accelerate massive particles with low charges. In 1996 Dana Andrews suggested that beam divergence might be reduced by neutralising the beam after acceleration and reducing residual beam-particle thermal motions.

Another approach to improving beam collimation was proposed by Gerald Nordley in 1999. It may become possible to nanoengineer beam particles so that they have enough intelligence and self-propulsion to 'home in' on the spacecraft.

7.8 BIBLIOGRAPHY

Andrews, D., 'Cost Considerations for Interstellar Missions', *Journal of the British Interplanetary Society*, **49**, 123–128 (1996).

Andrews, D. and Zubrin, R., 'Magnetic Sails and Interstellar Travel', *Journal of the British Interplanetary Society*, **43**, 265–272 (1990).

Belbruno, E. and Matloff, G. L., 'A Fast and Light Mission to Alpha/Proxima Centauri', in *Proceedings of 1993 AINA Conference – Advances in Nonlinear Astrodynamics*, ed. E. Belbruno, Geometry Center, University of Minnesota, Minneapolis, MN (November 8–9, 1993).

Bloomer, J. H., 'The Alpha Centauri Probe', in *Proceedings of the 17th International Astronautical Congress (Propulsion and Re-entry)*, Gordon and Breach, Philadelphia, PA (1967), pp. 225–232.

Cocks, J. C., Watkins, S. A., Cocks, F. H. and Sussingham, C., 'Applications for Deployed Superconducting Coils in Spacecraft Engineering: A Review and Analysis', *Journal of the British Interplanetary Society*, **50**, 479–484 (1997).

Driscoll, W. G. and Vaughan, W. (eds.), *Handbook of Optics*, McGraw-Hill, New York (1978), Chapter 8.

Forward, R. L., 'Zero Thrust Velocity Vector Control for Interstellar Probes: Lorentz Force Navigation and Circling', *AIAA Journal*, **2**, 885–889 (1964).

Forward, R. L., 'Round-Trip Interstellar Travel Using Laser-Pushed Lightsails', *Journal of Spacecraft and Rockets*, **21**, 187–195 (1984).

Forward, R. L., 'Starwisp: An Ultra-Light Interstellar Probe', *Journal of Spacecraft and Rockets*, **22**, 345–350 (1985).

Gilfillan Jr., E. S., *Migration to the Stars: Never Again Enough People*, Luce, Washington DC (1975).

Jackson IV, A. A. and Whitmire, D. P., 'A Laser-Powered Interstellar Rocket', *Journal of the British Interplanetary Society*, **32**, 335–337 (1978).

Jackson, J. D., *Classical Electrodynamics*, Wiley, New York (1962).

Jones, E., 'A Manned Interstellar Vessel Using Microwave Propulsion: A Dysonship', *Journal of the British Interplanetary Society*, **38**, 270–273 (1985).

Kare, J. T., 'Pulsed Laser Thermal Propulsion for Interstellar Precursor Missions', presented at STAIF 2000 Conference, University of New Mexico, Albuquerque, NM, January 30–February 3, 2000.

Korman, V., Gregory, D. A. and Peng, G., 'Optical Evaluation for Large Inflatable and Thin Film Fresnel Lens Components', presented at NASA/JPL/MSFC/AIAA Annual Tenth Advanced Space Propulsion Workshop, Huntsville, AL, April 5–8, 1999.

Landis, G. A., 'Beamed Energy Propulsion for Practical Interstellar Flight', *Journal of the British Interplanetary Society*, **52**, 420–423 (1999). Also see G. A. Landis, 'Dielectric Films for Solar and Laser-Pushed Lightsails', presented at STAIF 2000 Conference, University of New Mexico, Albuquerque, NM, January 30–February 3, 2000; and Landis, G. A., 'Dust Erosion of Interstellar Propulsion Systems', AIAA-2000-3339.

Matloff, G. L., 'An Approximate Heterochromatic Perforated Light-Sail Theory', IAA-95-IAA.4.1.01.

Matloff, G. L., 'Robosloth: A Slow Interstellar Thin-Film Robot', *Journal of the British Interplanetary Society*, **49**, 33–36 (1996).

Matloff, G. L., 'Use of Parabolic Solar Concentrators to Improve the Performance of an Interstellar Solar Sail', *Journal of the British Interplanetary Society*, 49, 21–22 (1996).

Matloff, G. L. and Fennelly, A. J., 'A Superconducting Ion Scoop and its Application to Interstellar Flight', *Journal of the British Interplanetary Society*, **27**, 663–673 (1974).

Matloff, G. L. and Potter, S., 'Near Term Possibilities for the Laser-Light Sail', in *Missions to the Outer Solar System and Beyond, 1st IAA Symposium on Realistic Near-Term Scientific Space Missions*, ed. G. Genta, Levrotto & Bella, Turin, Italy (1996), pp. 53–64.

Matloff, G. L., Walker, E. H. and Parks, K. 'Interstellar Solar Sailing: Application of Electrodynamic Turning', AIAA-91-2538.

Nordley, G. D., 'Momentum Transfer Particle Homing Algorithm', presented at NASA/JPL/MSFC/AIAA Annual Tenth Advanced Space Propulsion Workshop, Huntsville, AL, April 5–8, 1999.

Norem, P. C., 'Interstellar Travel: A Round Trip Propulsion System with Relativistic Velocity Capabilities', AAS 69–388.

Pikel'Ner, S. B., 'Structure and Dynamics of the Interstellar Medium', *Annual Review of Astronomy and Astrophysics*, **6**, 165–194 (1968).

Renk, K. F. and Genzel, L., 'Interference Filters and Fabry–Perot Interferometers for the Far Infrared', *Applied Optics*, **1**, 643–648 (1962).

Singer, C. F., 'Interstellar Propulsion Using a Pellet Stream for Momentum Transfer', *Journal of the British Interplanetary Society*, **33**, 107–116 (1980).

Vulpetti, G., 'A Critical Review on the Viability of Space Propulsion Based on the Solar Wind Momentum Flux', *Acta Astronautica*, **37**, 641–642 (1994).

Vulpetti, G. and Pecchioli, M., 'The Two-Sail Propulsion Concept', IAA-91-721.

Whitmire, D. P. and Jackson IV, A. A., 'Laser-Powered Interstellar Ramjet', *Journal of the British Interplanetary Society*, **30**, 223–226 (1977).

Wolfe, W. L., *Handbook of Military Infrared Technology*, US Office of Naval Research, Department of the Navy, Washington DC (1965).

Zubrin, R., *Entering Space*, Tarcher/Putnam, New York (1999).

Zubrin, R. and Andrews, D., 'Magnetic Sails and Interplanetary Travel', AIAA 89-2441.

8

On the technological horizon

It was a planet-sized shell of incandescence, where atoms were seized by its outer-most force-fringes and excited into thermal, fluorescent synchrotron radiation. And it came barely behind the wave front which announced its march. But the ship's luminosity was soon lost across light years. Her passage crawled through abysses which seemingly had no end.

Poul Anderson, *Tau Zero* (1970)

The early years of the space programme saw the origination of the interstellar ramjet. During the heady 1960s and 1970s, this approach seemed more than capable of permitting human exploration and colonsation of nearby solar systems. In fact, the entire Universe might open up to craft that followed the fictional track of Poul Anderson's *Leonora Christine*. More recently, the original ramjet concept has appeared less feasible, and we may have to content ourselves with its less capable derivatives. But regardless of our current opinion regarding its ultimate feasibility, the ramjet concept is too exciting to be abandoned. It may still lead to relativistic interstellar travel.

A number of variations on the ramjet concept are considered in this chapter. These include the original pure ramjet, the ram-augmented interstellar rocket, the laser ramjet and the ramjet runway. Also considered is the one proposed magnetic-scoop geometry that collects interstellar ions rather than reflecting them.

8.1 THE HYDROGEN-FUSING INTERSTELLAR RAMJET

The probable originator of the interstellar-ramjet concept is Robert Bussard, who published his provocative paper on the topic in 1960. An illustration of Bussard's proposal is presented in Figure 8.1.

A spacecraft equipped with a magnetic scoop (or ramscoop) moves through the interstellar medium. The scoop field collects interstellar ions (mostly protons), which move through a thermonuclear reactor capable of fusing protons to obtain helium,

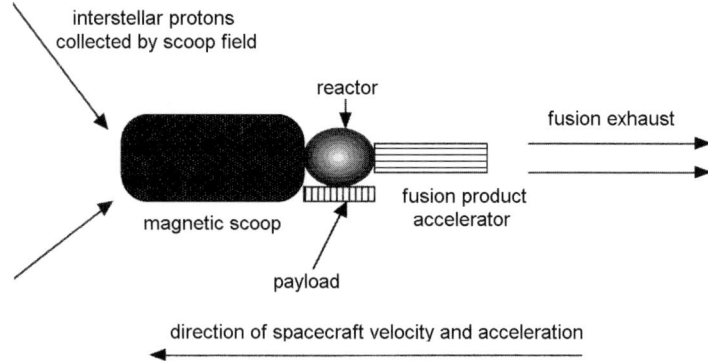

Fig. 8.1. The proton-fusing interstellar ramjet.

in a manner analogous to the reaction powering the Sun and most other main-sequence stars. Accelerated reaction products are exhausted out the spacecraft's rear, which produces thrust. The following describes a non-relativistic treatment of ramjet kinematics.

During each incremental time interval dt (in seconds), the mass of fuel collected by the ramjet's ramscoop can be written as

$$\frac{dM_f}{dt} = \rho_{\text{ion}} M_{\text{ion}} A_{\text{scoop}} V_s \text{ kg} \tag{8.1}$$

where ρ_{ion} is the density of the interstellar medium in ions per cubic metre, M_{ion} is the mass of a collected interstellar ion in kilogrammes, A_{scoop} is the scoop area in square metres, and V_s is the spacecraft velocity relative to the interstellar medium in metres per second.

If the ramscoop field does not impart a velocity component to the collected ions along the ship's line of flight, the ions arrive at the scoop with a velocity of V_s relative to the ship and are exhausted at $V_s + V_e$ relative to the ship, where V_e is the exhaust velocity. The previous discussion of nuclear rockets and Einstein's mass–energy conversion equation can be used to calculate V_e:

$$V_e = -V_s + \sqrt{V_s^2 + 2\Phi_{nf}\varepsilon_{nf}c^2} \text{ m s}^{-1} \tag{8.2}$$

where Φ_{nf} is the mass–energy conversion efficiency of the ship's fusion reactor, ε_{nf} is the efficiency of fuel transfer to the exhaust, and c is the speed of light in metres per second.

If we now consider momentum conservation at times t and $t + dt$ in a coordinate system moving with the craft at time t, $M_s(dV_s/dt) = V_e(dM_f/dt)$. Substituting equations (8.1) and (8.2), we obtain an expression for ramjet acceleration:

$$\frac{dV_s}{dt} = \frac{\rho_{\text{ion}} M_{\text{ion}} A_{\text{scoop}} c^2}{M_s}\left(-\beta_s^2 + \sqrt{\beta_s^4 + 2\Phi_{nf}\varepsilon_{nf}\beta_s^2}\right) \text{ m s}^{-2} \tag{8.3}$$

where $\beta_s = V_s/c$. Equation (8.3) is the basic equation for non-relativistic ramjet kinematics, and will be accurate for ship velocities less than about 0.2 c.

Exercise 8.1 Validate all intermediate steps in the derivation of equation (8.3).

In his original relativistic ramjet study, Bussard considered acceleration in Galactic nebulae with hydrogen densities of about 10^9 per m^3. Since most high-density nebulae consist of mostly neutral rather than ionised hydrogen, some form of fuel ionisation ahead of the ship would be required. As mentioned in a previous chapter, certain ramscoop magnetic fields might ionise neutral hydrogen. Another approach might be to project an ultraviolet laser beam to ionise hydrogen in front of the spacecraft, as proposed by Matloff and Fennelly in 1975.

But, as noted by Carl Sagan in 1963, the average interstellar hydrogen density is about 0.001 times the hydrogen density in dense star-forming Galactic nebulae. As mentioned in the last chapter, the local interstellar proton density is likely to be only 50,000 per m^3. The local density of neutral interstellar hydrogen is closer to 10^5 per m^3.

To evaluate ramjet performance using equation (8.3), consider a spacecraft with a mass of 10^6 kg moving at in initial velocity of 0.004 c through an interstellar medium with an ion density of 10^5 per m^3. The mass–energy conversion efficiency is 0.004, and 50% of the released nuclear energy is transferred to the exhaust. The ramscoop field has an effective radius of 1,000 km, and the mass of each collected interstellar proton is 1.67×10^{-27} kg. For this case, equation (8.3) becomes

$$\frac{dV_s}{dt} = 47.2\left(-\beta_s^2 + \sqrt{\beta_s^4 + 0.004\beta_s^2}\right) \text{ m s}^{-2} \quad (8.4)$$

Figure 8.2 presents the acceleration versus velocity profile for this ramjet between $\beta_s = 0.004$ and 0.1. For velocities below about 0.01 c, acceleration in m s^{-2} is approximately equal to $3\beta_s$. The increase of acceleration with velocity becomes less for higher velocities, approaching a value of about 0.095 m s^{-2} for velocities in excess of 0.1 c.

Between velocities of 0.004 c and 0.05 c, the average acceleration is about 0.045 m s^{-2}. The ramjet requires about about 9.7 years to accelerate between 0.004 c and 0.05 c. Since the average velocity of the ramjet is about 0.027 c during this time interval, the ship traverses about 0.26 light years during this time interval.

Between velocities of 0.05 c and 0.1 c, the ramjet's average acceleration is about 0.08 m s^{-2}. About six years are required to accelerate between 0.05 c and 0.1 c. At an average velocity of about 0.075 c, the craft traverses about 0.45 light years during this time interval.

Exercise 8.2 Apply the binomial series to expand the radical in equation (8.4) for the cases when the first term under the radical sign is very much smaller than the second term, and when the second term under the radical is larger than the first. After confirming that the discussion of low- and high-velocity accelerations for the above case are correct, apply the same approach to derive

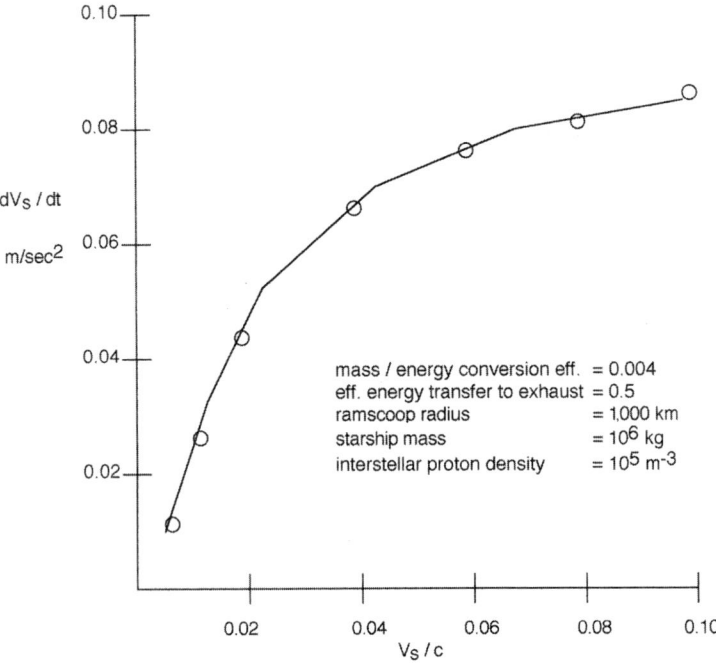

Fig. 8.2. Proton-fusing interstellar ramjet acceleration (m s^2) versus velocity/c.

low- and high-velocity non-relativistic ramjet accelerations for the general case defined by equation (8.3).

Unfortunately, a fusion reactor that could combine protons directly to obtain helium, in the manner of the Sun, seems to be hopelessly beyond technological capabilities. Stars hotter than the Sun obtain some or most of their fusion energy by the CNO cycle in which a carbon-12 nucleus reacts with a proton to form nitrogen-13 plus energy. After a series of reactions with protons in which various isotopes of carbon, nitrogen and oxygen are created, the final result is a carbon-12 nucleus, a helium-4 nucleus and energy. The carbon-12 nucleus is a catalyst, as it is not completely used, and it greatly speeds the reaction rate.

Although much easier to ignite than direct p–p fusion, the CNO chain may not be forever beyond the reach of fusion technology. In 1975, Daniel Whitmire analysed the problems and potentials of a CNO-cycle ramjet, and during the 1970s a number of researchers investigated the feasibility of a less capable version of the proton ramjet – the ram-augmented interstellar rocket.

8.2 THE RAM-AUGMENTED INTERSTELLAR ROCKET (RAIR)

In principle, RAIR (which was suggested by Alan Bond in 1974) overcomes the problems of obtaining energy from the interstellar medium by utilising interstellar

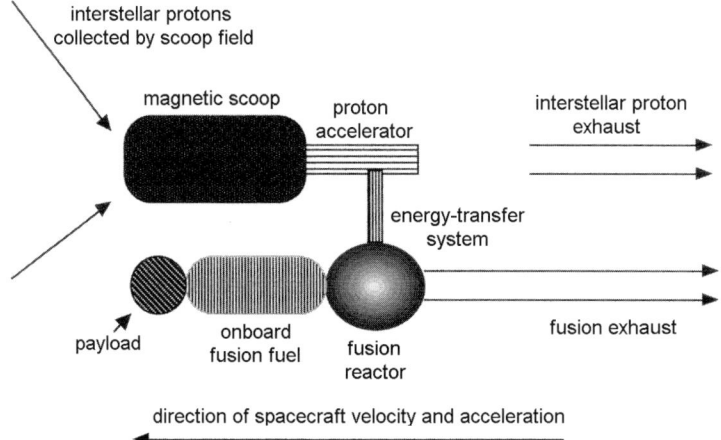

Fig. 8.3. A ram-augmented interstellar rocket.

matter as reaction mass and carrying an onboard supply of readily fusable isotopes. As presented in Figure 8.3, RAIR is essentially a two-stream thermonuclear rocket such as those considered in Chapter 6, except that the nuclear-inert fuel component consists of interstellar protons. The linear-momentum conservation equation for RAIR can be expressed as

$$M_s \left(\frac{dV_s}{dt} \right) = V_{e,nf} \left(\frac{dM_f}{dt} \right) + V_{e,if} \left(\frac{dM_{if}}{dt} \right) \tag{8.5}$$

where the subscript 's' denotes ship mass and velocity at time t, 'f' denotes onboard fuel mass, 'if' denotes interstellar ion fuel, and 'e' refers to velocities of the two exhaust streams relative to the ship. The mass of ion fuel collected per second – the second differential on the left-hand-side of equation (8.5) – is defined using equation (8.1).

We next assume that the rate of ion-fuel flow to onboard fuel flow is equal to a constant, $K_{\text{rair}} = dM_{if}/dM_f$. Assuming that all propulsive energy is obtained from the onboard fuel, the ion-fuel exhaust velocity can be defined as

$$V_{e,if}^2 + 2V_s V_{e,if} - 2U_{\text{rair}} = 0 \tag{8.6}$$

where $U_{\text{rair}} = c^2 \Phi_{ne} \varepsilon_{ne} \varepsilon_{ei} (K_{\text{rair}})^{-1}$, c is the speed of light, Φ_{ne} is the fraction of fusion fuel converted into energy, ε_{ne} is the efficiency of fusion-energy conversion into electric energy, and ε_{ei} is the efficiency of the ion-fuel linear accelerator.

High-speed RAIR approximation

We can solve equation (8.6) for $V_{e,if}$ using the quadratic formula if we define the condition $V_s^2 > 2U_{\text{rair}}$. Applying the binomial series, we find that $V_{e,if}$ is approximately equal to U_{rair}/V_s. Substituting this result and the definition for K_{rair} into equation (8.5), we obtain

$$M_s \left(\frac{dV_s}{dt} \right) = V_{e,nf} \left(\frac{dM_f}{dt} \right) + \frac{U_{rair} K_{rair}}{V_s} \left(\frac{dM_f}{dt} \right) \tag{8.7}$$

Remembering that $dM_s = -dM_f$, equation (8.7) can be rewritten for integration:

$$Ln \left(\frac{M_0 + M_f}{M_0} \right) = \int_{V_{in}}^{V_{fin}} \left(\frac{V_s}{V_{e,nf} V_s + U_{rair} K_{rair}} \right) dV_s \tag{8.8}$$

where M_0 and M_f are respectively masses of unfuelled RAIR and RAIR fusion fuel, and V_{in} and V_{fin} are respectively RAIR velocities at the beginning and end of high-speed RAIR operation. The right-hand side of equation (8.8) is a standard form, and can be easily solved. The term $(M_0 + M_f)/M_0$ is the mass ratio MR. Manipulating the solution, substituting for U_{rair} and K_{rair}, and applying the substitution $\beta = V/c$, we obtain the high-speed non-relativistic RAIR kinematics approximation:

$$\frac{MR_{rair}}{MR_{rocket}} = \left(\frac{\beta_{e,nf} \beta_{in} + \Phi_{ne} \varepsilon_{ne} \varepsilon_{ei}}{\beta_{e,nf} \beta_{fin} + \Phi_{ne} \varepsilon_{ne} \varepsilon_{ei}} \right)^{\varepsilon_{ne} \varepsilon_{ei}/(2\varepsilon_{nf})} \tag{8.9}$$

which compares RAIR and fusion-rocket mass ration for the case $V_s^2 > 2 U_{rair}$.

Low-speed RAIR approximation

To evaluate low-speed (*ls*) RAIR kinematics, we follow the 1976 arguments of Matloff. Assume that the spacecraft velocity relative to the interstellar medium is so low that we can ignore ion-flow velocity, and that both exhaust streams are well mixed. From the energy-conservation equation for RAIR exhaust, low-speed RAIR exhaust velocity can be approximated:

$$V_{e,rair,ls} \approx \sqrt{2 \Phi_{ne} \varepsilon_{ne} \varepsilon_{ei} c^2 \left(\frac{dM_{nf}}{dM_{if} + dM_{nf}} \right)} = V_{e,rock} \sqrt{\frac{1}{1 + K_{rair}}} \tag{8.10}$$

where all terms have been previously defined:
Since the momentum change of the ship during a small time interval is equal to the momentum change of the mixed ion and fusion fuels, and the incremental change in fusion-fuel mass is equal and opposite to the change in ship mass during a small time interval.

$$-\frac{dM_s}{M_s} \approx \frac{dV_s}{(\sqrt{1 + K_{rair}}) V_{e,nf}} \tag{8.11}$$

The low-speed RAIR mass ratio can now be approximated as a function of ship velocity increase during RAIR operation, nuclear-fuel exhaust velocity, and K_{rair}:

$$MR_{rair,ls} \approx \exp \left(\frac{\Delta V_s}{V_{e,nf} \sqrt{1 + K_{rair}}} \right) \tag{8.12}$$

Comparison with computer simulation: a sample RAIR mission

A number of simplifying assumptions were utilised in the above consideration of RAIR kinematics. First, we assumed non-relativisitic dynamics and that the interstellar ion fuel was nuclear inert. This is not necessarily true, since (as suggested by Bond) protons entering a shipboard reaction chamber at high energy could react with onboard fusion fuels such as lithium or boron. Although the proton–lithium and proton–boron reactions are relatively difficult to ignite and not as energetic as some of the other fusion alternatives, these reactions are aneutronic. A large fraction of the fusion energy, therefore, can be transferred to the exhaust streams.

The assumption was also made that K_{rair} – the ratio of interstellar to fusion fuel consumption – is a constant throughout RAIR operation. As discussed by Conley Powell during his 1975 and 1976 computer-optimisation studies of RAIR kinematics, K_{rair} can be optimised at each stage of RAIR operation.

In an operational RAIR mission, the first stage might be a solar sail to accelerate the spacecraft to about $0.004\,c$ relative to the interstellar medium. Instead, to compare with Powell's 1975 results, we assume that initial acceleration from planetary velocities is by the low-velocity RAIR approximation, followed by high-velocity RAIR acceleration to $0.134\,c$.

With Powell, we assume that $\Phi_{ne} = 0.002$, $\varepsilon_{nf} = 0.25$, and $\varepsilon_{ne}\varepsilon_{ei} = 0.25$. These assumptions imply that 25% of the fusion energy is transferred to rocket exhaust, 25% is transferred to ion exhaust, and 50% is radiated as waste heat. We also assume that $K_{rair} = 10$ throughout RAIR operation.

From the above discussion, the high-speed RAIR approximation applies if $V_s^2 > 2c^2\Phi_{ne}\varepsilon_{ne}\varepsilon_{ei}\,(K_{rair})^{-1}$. For this case, therefore, we assume that the low-speed RAIR-acceleration approximation applies for ship velocities below $0.02\,c$, and the high-speed approximation applies for higher ship velocities.

First, we apply equation (8.12) to estimate the low-speed RAIR mass ratio for acceleration to $0.02\,c$. The fusion-rocket exhaust velocity in this equation is $0.032\,c$. The RAIR mass ratio for acceleration from planetary velocities to $0.02\,c$ is 1.22, which compares well with Powell's result of 1.25.

Equation (8.9) was then applied to calculate (high-speed RAIR mass ratio)/(rocket mass ratio) for various values of β_{fin} and $\beta_{in} = 0.02$. Rocket mass ratios were also calculated for each velocity increment, and were used to estimate high-speed (*hs*) RAIR mass ratios. The total RAIR mass ratio for each velocity increment was calculated by multiplying the high-speed results by 1.22. Table 8.1 compares our results with Powell's (1975) results.

For velocities less than about $0.07\,c$, our results are in excellent agreement with Powell's results. Our RAIR mass ratios are about 10% higher than Powell's for ship velocities of about $0.09\,c$, and our approximation becomes invalid for velocities above $0.1\,c$.

Exercise 8.3 Powell reported that RAIR kinematics is very sensitive to small variations in subsystem efficiencies. To check this, repeat the above calculations for the case of a nuclear mass–energy conversion efficiency of 0.003, with 30% of the released energy transferred to each exhaust stream.

Table 8.1. Estimated RAIR mass ratios compared with Powell (1975) results

β_{fin}	MR_{rair}/MR_{rocket}	MR_{rocket}	$MR_{rair,hs}$	MR_{rair}	Powell's MR_{rair}
0.065	0.67	4.08	2.73	3.33	3.33
0.085	0.60	7.62	4.57	5.57	5.00
0.127	0.50	28.33	14.17	17.29	10.00
0.134	0.49	35.25	17.27	21.07	10.98

It is possible to utilise the data used to generate Table 8.1 to model various aspects of a RAIR mission. Assume first that RAIR operation commences at 0.004 c in an interstellar medium with 0.05 protons cm^{-3}. If the RAIR scoop has a radius of 2,000 km, equation (8.1) can be used to demonstrate that 0.0013 kg of interstellar protons are collected and accelerated each second. For $K_{rair} = 10$, about 13 grammes of fusion fuel are reacted each second when the ship's velocity is 0.004 c. When the ship's velocity has increased to a RAIR-shutoff velocity of 0.08 c, about 2.6 grammes per second of fusion fuel is consumed.

At 0.004 c, the ship's fusion reactor generates about 2.3×10^{10} W, half of which is waste heat. When the ship's velocity reaches 0.08 c, reactor energy production has increased to 4.7×10^{11} W.

A maximum of 2.35×10^{11} W of waste heat must be radiated by the ship's radiator subsystem. If this radiator has a maximum operational temperature of 2,100 K, a black-body radiation calculator can be applied to demonstrate that a radiator area of about 10^5 m^2 is required.

Assume next that the ship's unfuelled mass is about 10^6 kg, and the mass of fusion fuel is about 3×10^6 kg. For the fuel consumption rates considered here, many decades are required to accelerate from 0.004 c to 0.08 c.

It might be thought that one means of improving RAIR performance would be to simply increase the radius of the ramscoop by a factor of 2 or 5. But as our discussion of ramscoop design later in this chapter reveals, construction of even 2,000-km radius ramscoops will be very challenging.

But there are at least two other ways to improve performance of a ramjet or RAIR: the laser ramjet and the ramjet runway, which are considered in the following sections of this chapter.

8.3 THE LASER RAMJET

Figure 8.4 demonstrates the principles of the laser ramjet, first suggested by Whitmire and Jackson in 1977, and further analysed by Matloff and Mallove in 1988.

A laser power station is located close to the Sun so that its beam (with power P_{laser}) is transmitted to a laser receiver/converter mounted on the distant starship. The laser receiver/converter converts received laser light to electricity with an

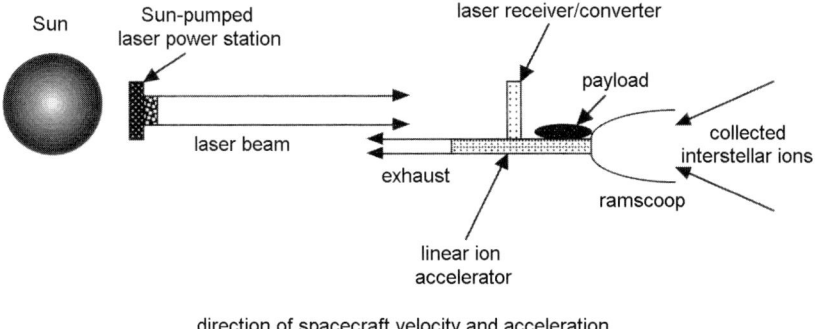

Fig. 8.4. A laser ramjet.

efficiency of ε_{lrc}. As with other ramjet versions, an electromagnetic scoop collects ions from the interstellar medium. These ions are accelerated by a linear accelerator and are emitted as exhaust. The energy/time transferred to the exhaust kinetic energy is $\varepsilon_{ei}\varepsilon_{lrc}P_{laser}$, where ε_{ei} is the efficiency of the ion-fuel linear accelerator.

Proceeding as before, the laser ramjet exhaust velocity can be written as

$$V_e = -V_s + \sqrt{V_s^2 + \frac{2\varepsilon_{ei}\varepsilon_{lrc}P_{laser}}{(dM_f/dt)}} \text{ m s}^{-1} \qquad (8.13)$$

where dM_f is the mass of interstellar ions collected in time interval dt, from equation (8.1).

Requiring conservation of momentum, and substituting for dM_f/dt, laser ramjet acceleration can be written as

$$\frac{dV_s}{dt} = \frac{\rho_{ion}M_{ion}A_{scoop}}{M_s}\left(-V_s^2 + \sqrt{V_s^4 + \frac{2\varepsilon_{ei}\varepsilon_{lrc}P_{laser}V_s}{\rho_{ion}m_{ion}A_{scoop}}}\right) \text{ m s}^{-2} \qquad (8.14)$$

Exercise 8.4 Demonstrate that both terms under the radical sign in equation (8.14) are dimensionally consistent.

Closed integration of equation (8.14) to analytically determine ship velocity as a function of time is daunting, at least for the velocities and laser powers of interest to early interstellar explorers. Instead of attempting such a task, Matloff and Mallove first defined the parameters of a reference mission in a typical interstellar medium, and then, from equation (8.14), calculated spacecraft acceleration as a function of velocity. These results were used to estimate the time required for a near-term laser ramjet.

For $M_s = 7.5 \times 10^5$ kg, $P_{laser} = 5 \times 10^{10}$ W, $\varepsilon_{ei}\varepsilon_{lrc} = 0.5$, a ramscoop radius of 1,000 km and an interstellar ion density of 0.05 cm^{-3}, the ramjet's acceleration is about 6×10^{-5} g. Varying laser power to maintain constant acceleration as velocity increases, about 80 years are required to accelerate from an initial velocity of $0.003\,c$ to $0.008\,c$. The laser power decreases to about 4×10^{10} W at the end of acceleration.

During the acceleration period, 0.46 light years are traversed. With deceleration by magsail, about 500 years would be required to reach α Centauri.

If the laser operates continuously until the start of magsail deceleration, one-way trips to α Centauri at 4.3 light years, τ Ceti at about 12 light years, and β Hydri at about 21 light years respectively require about 450, 600 and 750 years, allowing about 50 years for magsail deceleration.

The laser ramjet has an advantage over the laser light sail in that much less laser power is required. However, to achieve the full advantage of ramjet operation, it will be necessary to maintain laser-beam collimation for many decades or centuries. Perhaps thrustless turning could be applied to a laser ramjet to return it several times to a laser beam with a moderate collimation distance.

8.4 THE RAMJET RUNWAY

The final ramjet version to be considered in this chapter is the ramjet runway, presented schematically in Figure 8.5. First suggested by Whitmire and Jackson in 1977, and evaluated at non-relativistic velocities by Matloff in 1979, the ramjet runway is a compromise approach that compensates for the non-fusability of interstellar ions by preparing, in front of the spacecraft, a 'runway' consisting of electrically charged fusion micropellets. These micropellets would be deposited by a series of slow 'tanker' craft launched years or decades before the starship, using solar sails or electrical propulsion. The runway must be sufficiently collimated that even the modest ramscoops that we could currently design (such as the one discussed in the next section) could collect the fusion-fuel pellets and direct them into the onboard fusion reactor. As suggested by Gerald Nordley (in his 1999 paper cited in Chapter 7), nanotechnology might allow a degree of pellet intelligence so that the pellets could autonomously maintain good runway collimation prior to collection by the starship.

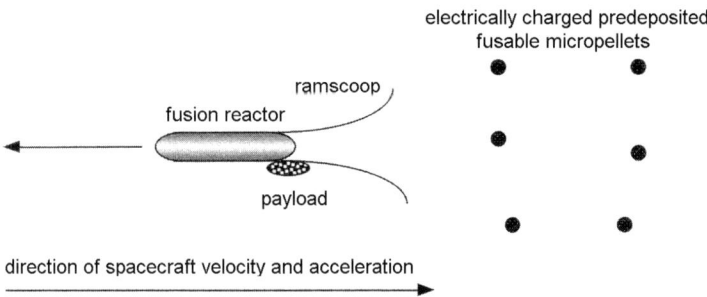

Fig. 8.5. The ramjet runway.

If the fuel pellets move at velocity V_{fp} relative to the interstellar medium, momentum-conservation arguments can be applied to obtain:

$$\frac{M_{fp}}{M_s} = \int_{V_{in}}^{V_{fin}} \frac{dV_s}{-(V_s - V_{fp}) + \sqrt{(V_s - V_{fp})^2 + 2\Phi_{ne}\varepsilon_{nf}c^2}} \tag{8.15}$$

where M_{fp} is the total mass of fusion fuel gathered by the ramscoop from the runway and the other parameters have been previously defined.

As demonstrated by Matloff (1979), this equation can be laboriously integrated applying trigonometric substitution. However, if the substitution $V_{sh} = V_s - V_{fp}$ is applied to equation (8.15), the equation can be very easily integrated for the high-velocity case: $V_{sh}^2 > 2\Phi_{ne}\varepsilon_{nf}c^2$. For high ship velocities relative to the interstellar medium,

$$\frac{M_{fp}}{M_s} = \frac{(\beta_{fin} - \beta_{fp})^2 - (\beta_{in} - \beta_{fp})^2}{\beta_{e,rock}^2} \tag{8.16}$$

where β_{fp} is V_{fp}/c, $\beta_{e,rock}$ is (exhaust velocity/c) for a fusion rocket with the same efficiencies as the ramjet's fusion propulsion system, and the other parameters have been previously defined.

Let us examine the case of a ramjet fusion propulsion system with $\Phi_{ne} = 0.004$ and $\varepsilon_{nf} = 0.15$. The velocity of the fusion runway relative to the interstellar medium is 0.004 c. From the condition for high-velocity operation, $V_{sh} > 0.0346c$. At the start of ramjet operation, $\beta_{in} = 0.0386$. For this case, $\beta_{e,rock}$ is 0.0346.

Table 8.2 presents fuel/unfuelled spacecraft masses for this ramjet and an equivalent fusion rocket ($M_{fp}/M_{s,rock}$). It is assumed that both craft use a solar sail for initial acceleration to 0.004 c, and that a fusion rocket with identical efficiencies is used to accelerate the ramjet from 0.004 c to 0.0386 c. The requirement for a fusion rocket will increase the fuel/unfuelled spacecraft mass ratios from equation (8.16), $M_{fp}/M_s(hs)$, by a factor of 1.72. Results for β_{fin} greater than 0.2 are not included in Table 8.2 because the non-relativistic approximation becomes less accurate at higher velocities.

Note that the ramjet runway's superiority over the fusion rocket increases as starship velocity increases. This is in agreement with the more accurate results in Matloff (1979).

Table 8.2. Performance of a fusion ramjet runway compared with a fusion rocket

β_{fin}	$M_{fp}/M_s(hs)$	$1.72 M_{fp}/M_s$	$M_{fp}/M_{s,rock}$
0.05	0.77	1.32	2.78
0.07	2.64	4.54	5.74
0.10	6.70	11.52	15.03
0.15	16.81	28.91	67.01
0.20	32.09	55.19	288.51

Consider a starship with an unfuelled mass of 10^7 kg, first accelerated to 0.004 c by a solar sail and then to 0.0346 c by the fusion rocket considered here. The total mass of the starship after launch from the Earth is 2.72×10^7 kg.

After rocket shut-down, the spacecraft enters a previously launched ramjet runway containing 6.7×10^7 kg of fusion micropellets receding from the Solar System at 0.004 c. The starship's velocity will be 0.1 c after traversing the runway.

If the runway has been prepared over a 50-year time interval prior to starship launch, the length of the runway will be 0.2 light years. At an average speed of 0.067 c, the starship traverses the runway in about three years. Its acceleration will be about 0.02 g. Such accelerations can be maintained by the toroidal-ramscoop structure discussed in the next section.

During every second of ramscoop operation, the ship ingests and fuses about 0.7 kg of fusion fuel. If all of the fuel reacts, the starship's fusion engine generates about 2.5×10^{14} W during acceleration – approximately ten times the total present-day terrestrial electrical energy consumption.

Exercise 8.5 Analyse a ramjet runway similar to the one considered in Table 8.2, except with a fuel pellet velocity of 0.002 c relative to the interstellar medium.

One alternative to the fusion rocket 'second stage' of the ramjet runway is to use a high-performance laser-rocket in which onboard propellant is energised by a laser beamed from the Solar System. Jackson and Whitmire analysed such a craft in 1978 (see Chapter 7).

Fission charges might be considered as fuel sources as an alternative to fusion micropellets along a ramjet runway. Such a possibility has been investigated in a recent paper by Roger Lenard and Ronald Lipinski.

The ramjet runway consideration above presumes that the fusion charges in the runway are ignited by either electron beams or laser beams. In 1995, M. I. Shmatov investigated ramjet runway kinematics in which head-on collisions between scooped particles ignite fusion microexplosions.

8.5 A TOROIDAL RAMSCOOP

In order to design a laser ramjet or ramjet runway, it is necessary to first demonstrate that some form of ramscoop is conceiveable, and that it will collect rather than reflect incoming interstellar protons or electrically-charged runway fuel pellets. As discussed in the previous chapter, the early solenoidal or current-loop ramscoop designs were more efficient at ion reflection than at ion collection. An electrostatic scheme of interstellar ion collection, proposed by Matloff and Fennelly in 1977, was based upon earlier studies of electrostatic interation of the interstellar plasma by Martin (1972), Langton (1973) and Whitmire (1975).

As pointed out by Cassenti in 1991, consideration of the interaction between the interstellar plasma and a moving electrostatically charged body is not a trivial problem in plasma physics. As discussed in the standard textbook by Jackson (1962),

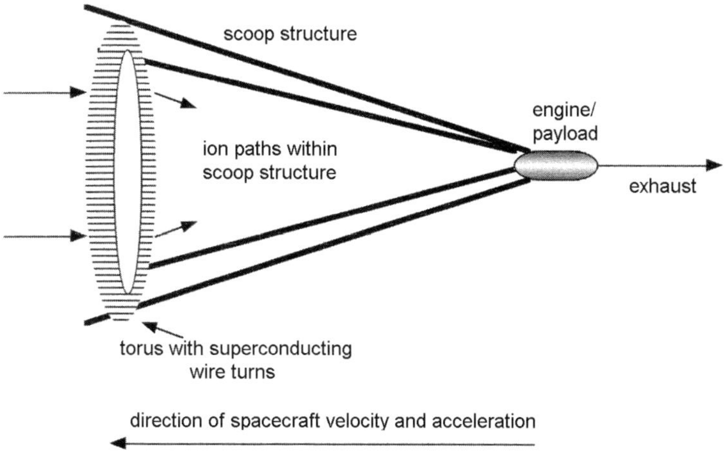

Fig. 8.6. The toroidal–magnetic ramscoop. Current flows along wires wound around the torus; positive ions are deflected inward.

Debye–Hückel screening might severely limit the performance of electrostatic interstellar-ion collection or deflection techniques. Interstellar ions from a great distance will tend to move in the direction of an oppositely charged electrostatic scoop. The charge of these ions will screen the scoop electrostatic field and reduce its intensity, from the point of view of more distant charges. Although spacecraft motion is a complicating factor, it seems likely that the effectiveness of electrostatic ion-collection techniques will be less than originally believed.

In 1991 Cassenti attempted to salvage the ramjet concept by proposing a toroidal-field ramscoop such as the one illustrated in Figure 8.6. Since this is an electromagnetic technique, Debye–Hückel screening need not apply. Because incoming ions are not affected by the scoop field until they are within the scoop, ion reflection will be minimal.

Many elementary physics texts, including Ohanian (1989), consider the operation of a magnetic torus such as the one shown in Figure 8.6. As shown in the figure, superconducting wire in a toroidal ramscoop is wound around the circumference of the torus. Magnetic field lines within the torus are closed circles. Depending upon current direction and ion charge, an ion entering the torus will be deflected either towards the centre or circumference of the doughnut-shaped torus.

From Ohanian, the magnetic field strength at a radial distance r_{torus} from the centre of a magnetic torus can be expressed as

$$B_{torus} = \frac{\mu_{fs} N_{wire} I_{wire}}{2\pi r_{torus}} \qquad (8.17)$$

where N_{wire} is the number of turns in the torus wire, I_{wire} is the current in the wire, and μ_{fs} is the permeability of the vacuum (1.26×10^{-6} Henry m^{-1}).

If the velocity of the spacecraft, V_s, is much greater than the relative velocity fluctuations of the interstellar gas, the force on an interstellar ion (F_{ion}) of charge q_{ion} entering the torus a distance r_{torus} from the torus centre will be

$$F_{ion} = V_s q_{ion} B_{torus} = \frac{V_s q_{ion} \mu_{fs} N_{wire} I_{wire}}{2\pi r_{torus}} \qquad (8.18)$$

If supercurrent direction and ion charge are such that the incoming interstellar ion is deflected towards the centre of the torus, the ion's acceleration under the influence of the ramscoop's magnetic field will be

$$acc_{ion} = \frac{F_{ion}}{m_{ion}} = \frac{V_s q_{ion} \mu_{fs} N_{wire} I_{wire}}{2\pi r_{torus} m_{ion}} \qquad (8.19)$$

where m_{ion} is the ion's mass. From elementary kinematics, the time for the ion to be deflected from r_{torus} to the torus centre will be

$$t_{ion} \cong \left(\frac{2 r_{ion}}{acc_{ion}}\right)^{1/2} \cong \left(\frac{4\pi r_{ion}^2 m_{ion}}{V_s q_{ion} \mu_{fs} N_{wire} I_{wire}}\right)^{1/2} \qquad (8.20)$$

Before they are focused at the scoop centre, the interstellar ions travel a longitudinal distance D_{ion} through the scoop. The distance can be estimated:

$$D_{ion} \cong V_s t_{ion} \cong \left(\frac{4\pi r_{ion}^2 m_{ion} V_s}{q_{ion} \mu_{fs} N_{wire} I_{wire}}\right)^{1/2} \qquad (8.21)$$

In his 1991 paper, Cassenti considered a toroidal scoop with a radius of 400 km, a supercurrent of 3×10^5 amps and twelve wire turns, travelling through the interstellar medium at 0 c. If interstellar protons ($q_{ion} = 1.6 \times 10^{-19}$ Coulombs, $m_{ion} = 1.67 \times 10^{-27}$ kg) are collected by this scoop, equation (8.21) reveals that the focus distance for ions entering about 200 km from the scoop centre is about 190 km. Cassenti, with a more accurate approximation, obtained an ion-focus length of about 170 km for this case.

Exercise 8.6 Estimate the focus length for deuterons (with twice the mass of protons and the same charge) entering the scoop at 0.01 c.

The variation of focal length with ionic mass/charge ratio may allow collection in the interstellar medium of fusable isotopes. As mentioned in the previous chapter, application of this or any other superconducting device in the inner Solar System to collect fusable ions from the solar wind (in the manner suggested by Matloff and Cassenti in 1992) will be limited by superconductor thermal constraints.

To maintain an electrically neutral scooped-in plasma, a toroidal ramscoop would be equipped with a low-current, oppositely-directed toroid to deflect interstellar electrons towards the scoop's centre. Grids could be used within the scoop interior to separate ions and electron flows.

As the starship accelerates, an unsupported wire structure in front of the craft would quickly collapse. Cassenti's scoop is stabilised with a combination of rotation-produced centripetal force and a supporting structure. Computer simulation reveals that accelerations of 0.04 g can be supported by the scoop structure. Electrical thrusters and radiation pressure from onboard lasers could also be applied to keep low-mass scoop components in position during starship acceleration. The total mass of Cassenti's ramscoop would be a few hundred thousand kilogrammes.

8.6 BIBLIOGRAPHY

Bond, A., 'The Potential Performance of the Ram-Augmented Interstellar Rocket', *Journal of the British Interplanetary Society*, **27**, 674–685 (1974).
Bussard, R. W., 'Galactic Matter and Interstellar Flight', *Astronautica Acta*, **6**, 179–194 (1960).
°Cassenti, B., 'Design Concepts for the Interstellar Ramjet', *Journal of the British Interplanetary Society*, **46**, 151–160 (1993); also published as AIAA 91-2537.
Jackson, J. D., *Classical Electrodynamics*, Wiley, New York (1962), p. 342.
Langton, N. H., 'The Erosion of Interstellar Drag Screens', *Journal of the British Interplanetary Society*, **26**, 481–484 (1973).
Lenard, R. X. and Lipinski, R. J., 'Interstellar Rendezvous Missions Employing Fission Propulsion Systems', presented at STAIF 2000 Conference, University of New Mexico, Albuquerque, NM, January 30–February 3, 2000.
Martin, A. R., 'The Effects of Drag on Relativistic Spaceflight', *Journal of the British Interplanetary Society*, **25**, 643–652 (1972).
Matloff, G. L. and Fennelly, A. J., 'Vacuum Ultraviolet Laser and Interstellar Flight', *Journal of the British Interplanetary Society*, **28**, 443 (1975).
Matloff, G. L., 'Utilization of O'Neill's Model I Lagrange Point Colony as an Interstellar Ark', *Journal of the British Interplanetary Society*, **29**, 775–785 (1976).
Matloff, G. L. and Fennelly, A. J., 'Interstellar Application and Limitations of Several Electrostatic/Electromagnetic Ion Collection Techniques', *Journal of the British Interplanetary Society*, **30**, 213–222 (1977).
Matloff, G. L., 'The Interstellar Ramjet Acceleration Runway', *Journal of the British Interplanetary Society*, **32**, 219–220 (1979).
Matloff, G. L. and Mallove, E. F., 'The Laser-Electric Ramjet: A Near Term Interstellar Propulsion Alternative', AIAA-88-3769.
Matloff, G. L. and Cassenti, B., 'The Solar Wind: A Source for Helium-3', IAA-92-228.
Ohanian, H. C., *Physics*, 2nd edn, Norton, New York (1989), p. 759.
Powell, C., 'Flight Dynamics of the Ram-Augmented Interstellar Rocket', *Journal of the British Interplanetary Society*, **28**, 553–562 (1975).
Powell, C., 'System Optimization for the Ram-Augmented Interstellar Rocket', *Journal of the British Interplanetary Society*, **29**, 136–142 (1976).
Powell, C., 'The Effect of Subsystem Inefficiencies Upon the Performance of the Ram-Augmented Interstellar Rocket', *Journal of the British Interplanetary Society*, **29**, 786–794 (1976).
Sagan, C., 'Direct Contact Among Galactic Civilizations by Relativistic Interstellar Spaceflight', *Planetary and Space Science*, **11**, 485–498 (1963).

Shmatov, M. I., 'Spacecraft Engine Based on Ignition of Microexplosions by Head-On Collisions', *Technical Physics Letters*, **21**, 155–156 (1995).

Whitmire, D., 'Relativistic Spaceflight and the Catalytic Nuclear Ramjet', *Acta Astronautica*, **2**, 497–509 (1975).

Whitmire, D. P. and Jackson IV, A. J., 'Laser Powered Interstellar Ramjet', *Journal of the British Interplanetary Society*, **30**, 223–226 (1977).

9

Exotic possibilities

'The time has come,' the walrus said,
'To talk of many things:
Of shoes—and ships—and sealing wax—
Of cabbages—and kings—
And why the sea is boiling hot—
And whether pigs have wings.'

Lewis Carroll, *Through the Looking-Glass* (1871)

A review of the preceding chapters will reveal that something is wrong with all approaches to interstellar flight that could be pursued using existing or foreseeable technology. Solar sails are feasible, but millennium-long travel times present problems for human science teams or peopled ships. Nuclear-pulse is technically feasible and perhaps a little faster, but how do you sell the world public on the prospect of storing large amounts of weapon-grade nuclear or thermonuclear material in near-Earth space during ship construction? Antimatter is technologically suitable and potentially very fast, but it is also very expensive. Only the slower ramjet alternatives such as the runway might prove feasible in the near term, and these might require many decades of preparation before a starship is launched. And the laser light sail (perhaps the current favourite, according to Frisbee and Leifer) requires not only the technical capability to beam a laser or maser over trillions of kilometres with a beam drift and accuracy measured in hundreds of kilometres, but also the continued terrestrial support for the mission during the decades-long or century-long acceleration process.

It is therefore not very surprising that scientists and engineers interested in travel to the stars have devoted some of their efforts to propulsion systems that seem at least as exotic as the walrus's 'cabbages and kings'. But analysts considering these options should be aware that many obstacles confront them – and not only the familiar obstacles of physics and technology.

Unfortunately, researchers in areas considered to be at or beyond the fringes of established science today must occasionally contend with the ridicule of their more

118 **Exotic possibilities** [Ch. 9

conventional colleagues. The reasons for this are certainly most complex, but have a great deal to do with increased competition for ever more elusive research funds.

The blame for the current state of affairs is certainly not one-sided. Those involved in 'breakthrough' or exotic research areas have sometimes allowed their enthusiasm to overtake their good judgement, and have publicly released research results before the normal peer-review journals have performed their function. Conservative and well-established researchers have, on the other hand, sometimes used heavy-handed tactics to suppress their more radical brethren and thereby reduce competition for research grants.

Sometimes this competition takes a humorous rather than a nasty turn. While serving as guest professor at the University of Siena, Italy, during the summer of 1994, I had the honour to meet an Italian physical chemist on the science faculty of that 800-year-old university, Prof Francesco Piantelli. Professor Piantelli's colleagues were impressed that he had obtained funding from the Italian Energy Board to attempt verification of the provocative and very controversial 1989 low-temperature nuclear-fusion results of Drs Stanley Pons and Martin Fleischmann (Mallove, 1989). However, many of his colleagues obviously enjoyed the fact that laboratory space in the ancient university is very limited, and that Piantelli's equipment had been moved to an unused wing of the Siena Psychiatric Facility.

In recent years, matters have become rather less volatile since NASA's Glenn (formerly Lewis) Research Center has been funding research potentially leading to space-propulsion breakthroughs. But breakthroughs causing major modifications of existing physical law may not be necessary to greatly increase our deep-space capabilities. Research performed during the summer of 1999 indicates that simple reinterpretation of undergraduate-level electrodynamics may yield positive results.

9.1 'SHOES AND SHIPS': THE POTENTIAL OF MAGNETIC SURFING

To reduce the amount of controversy surrounding breakthrough research, it would be an advantage to have some indication that such research might actually bear fruit. One method would be a simple demonstration that our concepts of physics – at least as far as they pertain to spacecraft propulsion – are less complete than commonly perceived. During the summer of 1999, while serving as a Faculty Fellow at the NASA Marshall Space Flight Center, I was part of a team that carried out such a demonstration. While application of the research requires a new astronomy rather than a new physics, and space-drive performance will never be equal to the more exotic approaches, magnetic-line surfing demonstrates that conventional propulsion scientists have overlooked at least one relevant implication of physical law: there may be many, many more surprises lurking just below the seemingly placid surface of conventional physics.

Select any college-level or university-level textbook that deals with electromagnetism, and proceed to the section on 'motional EMF' (often found in the chapter dealing with electromagnetic induction.) Figure 9.1(a) is derived from the figure that (apparently) invariably accompanies the text on this subject. It represents a

Sec. 9.1] 'Shoes and ships': the potential of magnetic surfing 119

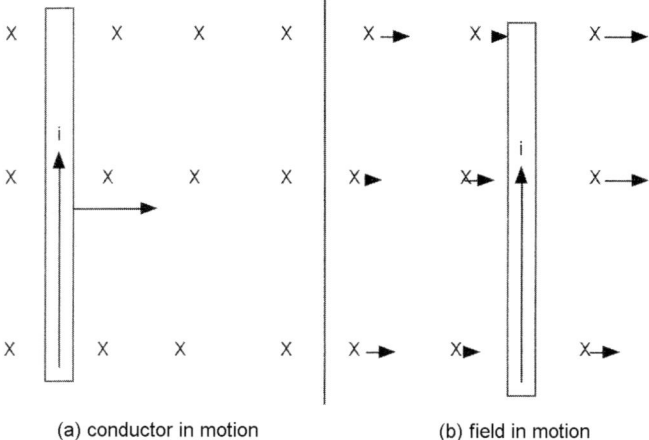

(a) conductor in motion (b) field in motion

Fig. 9.1. Current (i) produced in a conductor in motion relative to a magnetic field. Field lines (perpendicular to the page) are denoted by 'X'.

familiar application of Lenz's law. Consider an electrical conductor moving through a stationary magnetic field. An induced EMF (electromotive force) will be generated in the conductor to counter this motion. Essentially, if some of the electrical energy induced in the conductor is converted into useful work, the conductor slows down.

But what happens in the cosmically more interesting case where the magnetic field lines are moving more rapidly than the conductor, from the perspective of a terrestrial observer? The author and Les Johnson of NASA Marshall considered this question during the summer of 1999. Our supposition was that relative motion of the conductor and magnetic field, not absolute motion, was significant. For the case of a slowly moving conductor and a rapidly moving cosmic magnetic field (shown in Figure 9.1(b)), we felt that the process of converting electrical energy in the conductor into useful work will result in conductor deceleration *relative to* the moving magnetic field lines. From a terrestrial viewpoint, the conductor might accelerate dramatically, as if 'surfing' on the moving magnetic field lines.

To demonstrate the validity of our hypothesis, we asked NASA Marshall technician Bruce McCoy to supervise two summer student assistants at Marshall – Russell Lee and Alkesh Mehta, of New York City Technical College (CUNY) – in the construction of a simple double pendulum. As shown in Figure 9.2, the central arm of this pendulum consisted of an electrical conductor connected to a picoammeter to monitor the motion-induced EMF in the conductor. The outer arms consisted of a magnet, and the conductor and magnet could move together or independently. A null meter reading was obtained when the two pendulum arms were stationary or when they moved synchronously. But when either the magnet or the conductor were stationary and the other moved, current flowed in the conductor.

Devices that could be developed to exploit rapidly moving cosmic magnet fields for spacecraft propulsion must present the magnetic field with a unidirectional

120   **Exotic possibilities**   [Ch. 9

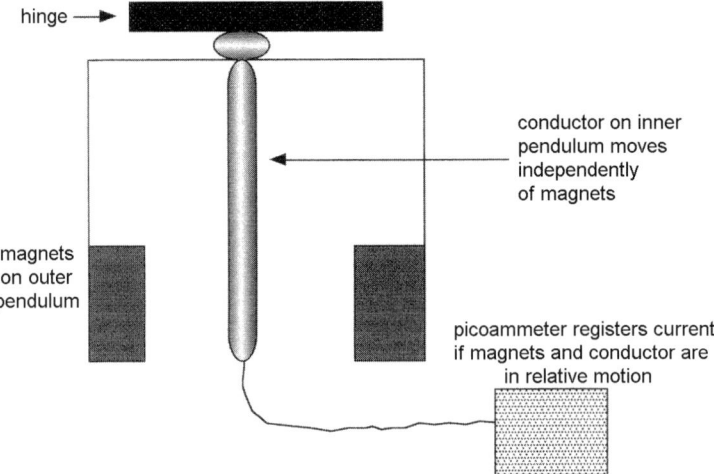

Fig. 9.2. A double pendulum designed to demonstrate currents induced by conductor motion relative to a magnetic field. The apparatus was constructed at NASA Marshall Space Flight Center during summer 1999.

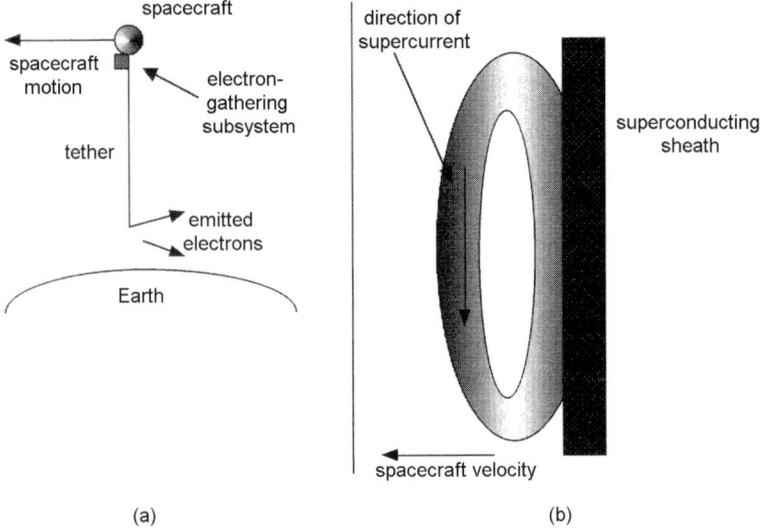

Fig. 9.3. Two methods of generating undirectional current flow in space. (a) An electrodynamic tether in Earth orbit; (b) a superconducting loop in which a superconducting layer shields current return from the interstellar magnetic field.

current. Two such devices – the electrodynamic tether and the partially sheathed superconducting 'wing' (or surfboard?) – are considered in a recent paper by Matloff and Johnson, and are presented in Figure 9.3. Electrodynamic tethers, as discussed by Ivan Bekey, have been tested successfully in near-Earth orbit. As shown in Figure

9.3(a), such tethers are deployed from satellites, and operate by moving through Earth's magnetic field while electrons are transferred between two layers in Earth's ionosphere. If the electrical energy of the electrons in the conducting tether is converted into useful work – to power a thruster, for example – the spacecraft decelerates relative to Earth's magnetic field. But NASA is considering a tether test in a magnetosphere of Jupiter (Gallagher, *et al.*, 1998). If Jupiter's magnetic field moves with the giant planet (which rotates every 10 hours), electrodynamic-tether operation near that planet may result in spacecraft acceleration, from the terrestrial viewpoint.

The velocity of the interstellar magnetic field lines relative to the Sun in its orbit around our Galaxy's centre is a matter of conjecture. If Galactic field lines are 'frozen into' local flows of ionised gases, the relative velocity will probably be less than a few hundred kilometres per second, and interstellar magnetic surfing will be unlikely.

However, what if a component of the local Galactic field is due to rapidly rotating sources such as neutron stars (Zhang *et al.*, 1998) or black holes (Zhang *et al.*, 1997)? Then, an interstellar spacecraft might surf the moving field lines using a tether or the partially shielded superconducting loop shown in Figure 9.3(b). In such a device, the Galactic field lines 'see' a unidirectional current because the current return is sheathed by layers of superconducting material that magnetic field lines cannot penetrate.

In such a rapidly moving Galactic magnetic field, a starship might first exit the Solar System using a solar sail, after which it could deploy one of the magnetic devices in Figure 9.3 or an equivalent system to 'surf' to the velocity of the magnetic field lines. The supercurrent loop could then be reconfigured (as discussed in Chapter 7) to complete a thrustless turn to the desired trajectory.

It is impossible to estimate the performance of an interstellar magnetic surfer without *in situ* measurements of the near interstellar magnetic field. Perhaps the early heliopause probes of the next few decades could be equipped with a tether-based magnetometer to estimate the velocity of the local interstellar magnetic field relative to the Sun.

A different type of magnetic 'shoe' – the mini-magnetosphere – has been suggested by Robert Winglee (2000) of the University of Washington, and mentioned in the literature by Robert Cassanova of the NASA Institute for Advanced Concepts. In this approach, a magnetic field is generated around the spacecraft with a configuration similar to the Earth's magnetic field. Like the Earth's field, the mini-magnetosphere could deflect the incoming plasma. If this proves feasible, and if superconductors are not required, propulsion by the reflection of the solar wind, magnetic shielding from cosmic rays and other inner Solar System magnetic-field applications will become possible.

A final method of increasing the efficiency of interstellar spacecraft, as suggested by Clint Seward of Electron Power Systems Inc., might be to store energy as magnetic field energy within a hollow torus around which electrons spiral. If the idea proves to be feasible, propellant could be expelled at tremendous velocities after being heated by elastic collisions with the torus surface.

9.2 'SEALING WAX': APPROACHES TO ANTIGRAVITY

The 'sealing wax' (to paraphrase Lewis Carroll's walrus once again) of many a science fiction epic is antigravity. As most readers know, a hypothetical antigravity machine generates some form of field in which the direction of gravitational lines of force is reversed. Instead of being an attractive force such as gravity, antigravity repulses. Might such a thing actually become possible one day in the real world, or will antigravity always be a favourite of the science fiction epic? This is an impossible question to answer at present. A number of physicists – notably Huseyin Yilmaz – have developed alternatives to general relativity theory in which antigravity is allowed. But it is difficult for such alternatives to succeed when considering the success of general relativity in explaining such phenomena as the advance of Mercury's perihelion, the existence of gravitational lenses, and the observed variation of star positions near the Sun during a total solar eclipse.

Some experimentalists, including David Noever and Ron Koczor, have investigated possible couplings between electromagnetism and gravity that might one day be used to produce antigravity. According to the experimental results, a small change in the mass (less than 1 part in 1,000) of a test mass suspended above a rotating superconductor might occur. But these experiments are difficult to perform, and many physicists question whether the results are replicable. (See Alexandre Szames' monograph to review other antigravity concepts.)

9.3 'A BOILING-HOT SEA': ZERO-POINT ENERGY AND SPECIAL-RELATIVISTIC STAR DRIVES

As discussed by Henning Genz, the physical Universe seems to result from a stabilised quantum-level fluctuation in the underlying 'foam' of the universal vacuum. Empty space, at distances of about 10^{-35} m, is a dynamic, 'boiling-hot sea' of particles constantly coming into and going out of existence, and enormous 'positive' energies exactly balanced by their 'negative' energy counterparts (so that the observed time-averaged effects are exactly zero).

Something happened in the first instants of the Universe to stabilise a vacuum fluctuation and produce our spacetime. Could we perhaps duplicate this effect on a much smaller scale and use the resulting vacuum of zero-point energy (ZPE) to propel a starship (as suggested by Robert Forward and David Froning)?

As mentioned by Genz, ZPE research is not without risks. If we attempt to extract vacuum energy by duplicating the conditions of the Big Bang, we may inadvertently erase our Universe in a brand-new Creation. Most people would consider such an event a major environmental impact.

But such extreme energy and field levels may not be necessary. Forward discusses the work of H. B. G. Casimir during the 1940s. Casimir investigated the distance variation of the short-range electromagnetic interaction between molecules, and predicted what has come to be known as the Casimir effect – that a portion of the force between the two conducting objects with a very small separation is due to

Sec. 9.3]   'A boiling-hot sea': zero-point energy and special-relativistic star drives   123

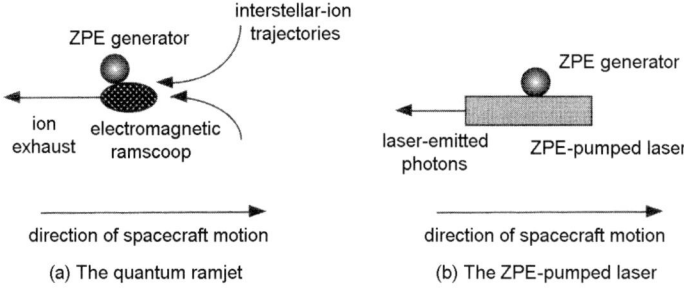

Fig. 9.4. Two hypothetical ZPE-propelled starships.

zero-point fluctuations of the vacuum. Specifically, any pair of conducting plates with a small enough separation (less than 1 µm) will experience a non-electrostatic force component due to the fact that some of the vacuum fluctuations do not exist in the small space between the conducting plates. For long a theoretical curiosity, the Casimir effect was finally confirmed by Steven Lamoreaux in 1997.

Forward presented a conceptual vacuum-fluctuation battery in which a large number of similarly charged ultrathin metallic leaves or a flat spiral of ultrathin metallic sheet are arranged in a stack with separations in the order of 1 µm. By appropriately adjusting the plate separation, it may be possible to obtain net ZPE energy for propulsive purposes. Casimir micromachines have also been discussed by Jordan Maclay.

A more elegant approach would be to obtain ZPE from some physical process that would require fewer moving parts. One possibility is sonoluminescence, a phenomenon in which high-frequency light is emitted by the sound-driven expansion or contraction of gas bubbles in fluids. According to Claudia Eberlein of the University of Illinois, one theoretical explanation for this mysterious phenomenon is ZPE produced by a type of Casimir effect. However, James Glanz has discussed alternative theoretical explanations for sonoluminescence that do not require ZPE.

If it proves possible to obtain large quantities of energy from the vacuum, a number of relativistic space drives will become possible. Froning has suggested a quantum ramjet (Figure 9.4(a)) that obtains reaction mass from the interstellar medium (in the same manner as the interstellar ramjets discussed in the previous chapter). The energy to accelerate the interstellar ions would come from a ZPE machine rather than a fusion reactor.

Matloff (1997) proposed an alternative approach (Figure 9.4(b)) in which ZPE is used to 'pump' an onboard laser, and the spacecraft is propelled by the momentum of the laser-emitted photons. Although less effective than the quantum ramjet because ions have more momentum than photons, the ZPE-laser may be capable of relativistic velocities if the ZPE machine's specific power (in kilowatts per kilogramme) and the laser efficiency are high enough.

But there may be more subtle ways to use vacuum fluctuations to allow a spacecraft to approach the speed of light. Harold Puthoff and Bernard Haisch have suggested that inertia might be caused by an interaction between the 'real' particles of matter and the 'virtual' particles of the universal vacuum. If this is indeed true, it is not impossible that we may one day learn how to 'polarise' the vacuum in front of a starship and thereby greatly reduce its inertia.

If some form of ZPE propulsion becomes possible, space travel at speeds close to the speed of light may become practicable. Starship designers will be compelled to include the effects of Einstein's special theory of relativity in the mission plans.

Special relativity and high-speed spacecraft kinematics

One of the basic principles of special relativity theory is that the speed of light, c, is a constant to all observers, regardless of observer or light-source velocity, and that c is the basic speed limit of the Universe. If we define measurements of a spacecraft's mass (M_s), length (L_s) and elapsed time (t_s) in an unaccelerated or Galactic reference frame (grf) and an accelerated spacecraft reference frame (srf) for a spacecraft velocity of V_s, we can compare measurements of these parameters in the two reference frames using the Lorentz parameter, γ (Fowles, 1962):

$$M_{s,grf} = M_{s,srf}\gamma, \, L_{s,grf} = L_{s,srf}\gamma^{-1}, t_{s,grf} = t_{s,srf}\gamma \quad (9.1)$$

where $\gamma = [1 - (V_s/c)^2]^{-1/2}$.

Let us examine the case of a spacecraft with a design mass of 10^6 kg and a design length of 100 m passing an unaccelerated observer at a velocity of 0.5 c. For such a situation, $\gamma = 1.15$. The unaccelerated observer will measure the spacecraft's mass and length respectively as 1.15×10^6 kg and 87 m. A time interval measured as one hour onboard the spacecraft will have a duration measured by the unaccelerated observer of about 69 minutes.

> *Exercise 9.1* Calculate γ and compare ship masses, lengths and elapsed times from the point of view of the unaccelerated and accelerated observers for the ship described above and $V_s = 0.1$ c and 0.9 c.

From the pont of view of the Galactic observer, the accelerated ship's mass will approach infinity as its velocity approaches the speed of light. Simultaneously, the ship's length will approach zero and the ship's onboard clocks will seem to stop, as measured by the unaccelerated observer.

There have been many confirmations of these bizarre effects – notably, the mass/energy conversion equation that results in the release of nuclear energy. Also, atomic accelerators are designed to compensate for the increase of particle mass with acceleration, and accelerated radioactive species take longer to decay, following the predictions of special relativity.

From special relativity alone, it might be suspected that the effects are symmetrical: a shipbound observer might see the same effects in the unaccelerated reference frame he is passing. As described by Ohanian (1989), an experiment was performed in the 1980s to determine whether time dilation – the slowing down of accelerated clocks – really occurs. Two highly accurate atomic clocks were calibrated on the ground. One remained in place as its twin flew around the world aboard a commercial jetliner. When the two clocks were later compared, the airborne clock had slowed, as predicted by special relativity.

Additional relativistic transformations generalise the linear momentum ($M_s V_s$ for a ship of mass M_s moving at a non-relativistic velocity of V_s) and the kinetic energy ($\frac{1}{2} M_s V_s^2$ for a non-relativistic starship), for the case of relativistic velocities. From the point of view of an observer in the Galactic reference frame, the linear momentum and kinetic energy of a starship moving at a near-optic velocity are (Sears et al., 1977)

$$P_{s,grf} = M_{s,srf} \gamma V_s, \quad KE_{s,grf} = M_{s,srf} c^2 (\gamma - 1) \tag{9.2}$$

Exercise 9.2 In the expression for relativistic kinetic energy in equation (9.2), first expand γ using the binomial theorem, and confirm that the relativistic kinetic equation reduces to the Newtonian kinetic equation for low velocities. Then plot the ratio of relativistic to Newtonian kinetic energies versus V_s for β_s between 0.1 and 0.9.

Many authors (including Marx, Shepherd, and Oliver) have applied the above relativistic transformations to the case of the relativistic rocket. An expression for the mass ratio of a relativistic rocket has been derived:

$$M.R._{\text{rel}} = \left(\frac{1 + \beta_s}{1 - \beta_s} \right)^{c/(2V_e)} \tag{9.3}$$

We can compare relativistic with Newtonian mass ratios calculated using equation (4.4). Taken an exhaust velocity of 0.1 c and values of $\beta_s (V_s/c)$ of 0.1 and 0.2. At $\beta_s = 0.1$, both equations yield nearly equal mass ratios of 2.72. For $\beta_s = 0.2$, the relativistic mass ratio is 7.59 and the Newtonian is 7.2.

Exercise 9.3 As velocity increases, the relativistic rocket mass ratios become greater than the corresponding Newtonian mass ratios. Check this by calculating mass ratios for $\beta_s = 0.3$ and 0.4.

In his 1980 paper on the quantum ramjet, Froning investigated the relativistic kinematics of this spacecraft in its own accelerated reference frame (*srf*). From Froning's equation (6), quantum ramjet acceleration in its own reference frame can be approximated:

$$\frac{dV_{qr,srf}}{dt} \cong \frac{c^2 \rho_{in,grf} A_{scoop,grf}}{M_{s,srf}} f_{qr} \Phi_{nf} \varepsilon_{nf} \tag{9.4}$$

where $\rho_{in,grf}$ is the interstellar ion mass density in the Galactic frame of reference, $A_{scoop,grf}$ is the ramscoop's field area in the Galactic frame of reference, $M_{s,srf}$ is the ship's mass in its accelerated reference frame, Φ_{nf} is fusion efficiency of mass–energy conversion, ε_{nf} is the efficiency of energy transfer to the spacecraft ion exhaust stream, and f_{qr} is the ratio of quantum energy per ion to energy per ion released from a fusion reactor.

The non-relativistic performance of a ZPE-laser is discussed in Matloff's 1997 paper. The Newtonian acceleration of a ZPE-laser can be expressed as $P_{\text{laser}}/(M_s c)$, where P_{laser} is laser power, M_s is ship mass, and c is the speed of light. A relativistic treatment of this spacecraft's kinematics has not yet been performed.

9.4 'CABBAGES AND KINGS': GENERAL RELATIVITY AND SPACETIME WARPS

General relativistic effects cannot be ignored in really accurate treatments of even slow interstellar travel because, as Banfi has pointed out, an inner Solar System starting point for any interstellar expedition is within the warped spacetime produced by the Sun's gravitational field. But such refinements might be considered the 'cabbages' of general relativity. The 'kings' of the general relativistic theory are, of course, the spacetime warps long used as universal shortcuts by science fiction writers. Space-warp concepts applied to interstellar space travel have been reviewed by Forward in 1985, Matloff in 1996, and Cassenti/Ringermacher in 2000.

While it is true that an object of solar mass and size warps the fabric of spacetime in its vicinity, much more extreme conditions are required to construct a time machine or a space drive capable of reaching the stars in times much less than a human lifetime. To manifest such conditions, it is necessary to create a 'singularity' – a place where the spacetime curvature approaches infinity (Kaufmann, 1979). To do this gravitationally, the singularity creator must construct a 'black hole'.

Million-solar-mass black holes left over from the early epochs of the Universe apparently reside in the centre of spiral galaxies such as our Milky Way. Smaller star-mass black holes are produced as a final stage in the evolution of very massive stars. A black hole forms when the collapsar's gravitational energy approaches its total mass energy (see Eardley and Press, 1975).

At the event horizon of a black hole, the mass-density of the collapsed celestial object (collapsar) is so great that light cannot escape. Substituting the speed of light, c, for the escape velocity of a collapsar, we can easily calculate the Schwartzchild radius of the singularity:

$$R_{sch} \cong \frac{2GM_{coll}}{c} \quad (9.5)$$

where G is the gravitational constant, and M_{coll} is the collapsar's mass. Objects within the event horizon of a black hole have effectively departed from our Universe. A solar-mass collapsar has a Schwartzschild radius of about 1.5 km.

Exercise 9.4 First calculate the Schwartzchild radius of the 10^6-solar-mass black hole suspected of lurking at the centre of the Milky Way Galaxy. Then test the hypothesis that the entire Universe may be a black hole. Do this by calculating the Schwartzchild radius of a universe containing 3×10^{23} solar-mass stars, and by confirming that this radius approximates tens of billions of light years.

If we can locate a handy collapsar, we could conceiveably use it to leave normal spacetime and take a shortcut through some higher dimensional hyperspace (a 'worm hole'), provided, that is, that we can locate an appropriate aperture to normal space (often called a 'white hole'), understand the intricacies of hyperspatial travel, and overcome such inconveniences as the enormous tidal stresses experienced as we approach the Schwartzschild radius. But these problems might seem to be somewhat academic, since no black holes have been detected within the cosmic vicinity of our Solar System.

One author who has considered how we might alleviate this apparent shortage of nearby stellar-mass black holes is Adrian Berry, who suggested in 1977 that interstellar ramjet/magsail technology (see Chapters 7 and 8) might be used to create an artificial stellar-mass singularity. Berry suggests that an enormous fleet of ramjets could 'herd' the interstellar medium over a vast volume of space near the Solar System, so that huge quantities of interstellar gas could be induced to converge on the same point.

Somewhat more immediate are the suggestions that we might create an artificial singularity using means other than gravity. Miguel Alcubierre and Yoshinari Minami have independently suggested that we might do this using magnetic fields many orders of magnitude greater than those produced on the Earth – even greater than those at the surface of a neutron star or exotic fields that might be manifested from the universal vacuum. Alcubierre's and Minami's ships (if possible) would be pushed or pulled through the Universe by a bubble of warped spacetime.

A good deal of research is still required before we can determine the ultimate feasibility of warp drives. Some current research in this area is being funded through the NASA Breakthrough Propulsion Office, directed by Marc Millis of NASA Glenn (formerly Lewis) Research Center. One goal of this research is to create a short-lived mini-singularity in the laboratory in order to conduct *experimental* general relativity research. Maccone, Davis and Landis have investigated the possibility of constructing a magnetically warped region of spacetime that would satisfy the Levi–Civita solution of Einstein's general relativistic equation.

The requisite (gigagauss) magnetic fields could be generated for a tiny fraction of a second by a new generation of pulsed lasers. As described by Mourou *et al.*, Perry, and Mourou and Perry, these devices have an energy in the vicinity of 1,000 J. Because their pulse duration is about 10^{-12} s, pulse powers of 10^{15} W are possible. These are table-top devices with costs approximating $500,000 (US).

Breakthrough-propulsion researchers would hope that some type of spacetime hysteresis effect would allow their mini-singularity to exist for less more than 10^{12} s. But are there any risks to this research?

In September 1999, author/journalist Fred Moody discussed, on a Web site (www.ABCNEWS.com), his concerns about performing mini-black-hole research in terrestrial laboratories. Although it is very improbable, it is not entirely impossible that a laboratory-generated black hole could survive longer than expected, grow enormously, and even threaten the Earth. As is the case with ZPE researchers, black-hole experimentalists must check their predictions very, very carefully before attempting to produce their mini-singularities.

9.5 'WINGED PIGS': SOME OTHER EXOTIC IDEAS

Many other non-conventional approaches to interstellar travel have been suggested. Here, we consider a few of them.

In 1997, George Miley – a respected plasma physicist from the University of Illinois – reviewed the evidence for anomalous energy produced when hydrogen and deuterium atoms are loaded into solid lattices at room temperature. Originally called 'cold fusion', such anomolous-energy results indicate that our understanding of low-temperature nuclear and thermonuclear reactions may be incomplete. When and if a successful theory of these effects emerges, significant applications for deep-space propulsion may become apparent.

In 1977, W. E. Moeckel suggested that a 'thrust-sheet' of radioisotopes or fissionable materials could produce high specific impulses if isotope decay or fission products could be emitted in a highly directional manner. It may one day be possible – as suggested by J. A. Morgan in 1999 – to utilise directional neutrino emission in the same manner.

We should keep an open mind about all these ideas. Although 99% of the Breakthrough Propulsion suggestions may be crazy, the other 1% may be gems!

9.6 BIBLIOGRAPHY

Alcubierre, M., 'The Warp Drive: Hyper-Fast Travel Within General Relativity', *Classical Quantum Gravitation*, **11**, L73–L77 (1994).

Banfi, V., 'Theory of Relativistic Flight to Oort's Comet Cloud', in *Missions to the Outer Solar System and Beyond, 2nd IAA Symposium on Realistic Near-Term Scientific Space Missions*, ed. G. Genta, Leviotto & Bella, Turin, Italy (1998), pp. 31–36.

Bekey, I., 'Tethers Open New Space Options', *Astronautics and Aeronautics*, 32–40 (April 1983).

Berry, A., *The Iron Sun*, Warner, New York (1977).

Casimir, H. G. B., 'On the Attraction Between Two Perfectly Conducting Plates', *Proceedings Koninklijke Nederlandse Akademie van Wetenshappen, Amsterdam*, **51**, 793–796 (1948).

Cassenti, B. N. and Ringermacher, H. L., 'Engineering Warp Drives', presented at STAIF 2000 Conference, University of New Mexico, Albuquerque, NM, January 30–February 3, 2000.

Cassanova, R. A., 'Overview: NASA Institute of Advanced Concepts', presented at NASA/JPL/MSFC/AIAA Annual Tenth Advanced Space Propulsion Workshop, Huntsville, AL, April 5–8, 1999.

Davis, E., 'Wormhole Induction Propulsion (WHIP)', in *NASA Breakthrough Propulsion Physics Workshop Proceedings, NASA/CP-1999-208694*, ed. M. Millis, NASA Lewis (Glenn) Research Center, August 12–14, 1997.

Eardley, D. M. and Press, W. H., 'Astrophysical Processes Near Black Holes', *Annual Reviews of Astronomy and Astrophysics*, **13**, 381–423 (1975).

Eberlein, C., 'Sonoluminescence as Quantum Vacuum Radiation', *Physical Review Letters*, **76**, 3842–3845 (1996).
Eberlein, C., 'Theory of Quantum Radiation Observed as Sonoluminescence', *Physical Review A*, **53**, 2772–2787 (1996).
Forward, R. L., 'Alternate Propulsion Energy Sources', AFRPL-83-067 (1983).
Forward, R. L., 'Extracting Electrical Energy from the Vacuum by Cohesian of Charged Foliated Conductors', *Physical Review B*, **30**, 1700–1702 (1984).
Forward, R. L., 'Space Warps: A Review of one Form of Propulsionless Transport', *Journal of the British Interplanetary Society*, **42**, 533–542 (1989).
Fowles, G. R., *Analytical Mechanics*, Holt, Rinehart and Winston, New York (1962).
Frisbee, R. H. and Leifer, S. D., 'Evaluation of Propulsion Options for Interstellar Mission', AIAA 98-3403.
Froning, H. D., 'Propulsion Requirements for a Quantum Interstellar Ramjet', *Journal of the British Interplanetary Society*, **33**, 265–270 (1980).
Froning, H. D., 'Investigation of a Quantum Ramjet for Interstellar Flight', AIAA 81-1533.
Gallagher, D. L., Johnson, L., Moore, J. and Bagenal, F., 'Electrodynamic Tether Propulsion and Power Generation at Jupiter', NASA Technical Publication TP-1998-208475 (June 1998).
Genz, H., *Nothingness: The Science of Empty Space*, Perseus, Reading, MA (1999).
Glanz, J., 'The Spell of Sonoluminescence', *Science*, **274**, 718–719 (1996).
Haisch, B., 'The Zero-Point Field and the NASA Challenge to Create the Space Drive', in *NASA Breakthrough Propulsion Physics Workshop Proceedings, NASA/CP-1999-208694*, ed. M. Millis, NASA Lewis (Glenn) Research Center, August 12–14, 1997.
Kaufmann III, W. J., 'Black Holes and Warped Spacetime', Freeman, San Francisco, CA (1979).
Lamoreaux, S. K., 'Demonstration of the Casimir Force in the 0.6 to 6 μm Range', *Physical Review Letters*, **78**, 5–8 (1997).
Landis, G., ''Magnetic Wormholes' and the Levi-Civita solution to the Einstein Equation', *Journal of the British Interplanetary Society*, **50**, 155–157 (1997).
Maccone, C., 'Interstellar Travel Through Magnetic Wormholes', *Journal of the British Interplanetary Society*, **48**, 453–458 (1995).
Jordan Maclay, G., 'A Design Manual for Micromachines Using Casimire Forces: Preliminary Considerations', presented at STAIF 2000 Conference, University of New Mexico, Albuquerque, NM, January 30–February 3, 2000.
Mallove, E. F., *Fire from Ice*, Wiley, New York (1991).
Matloff, G. L., 'Wormholes and Hyperdrives', *Mercury: The Journal of the Astronomical Society of the Pacific*, **25**, No. 4, 10–14 (July–August 1996). (*Mercury* editor G. Musser contributed to this paper, which also contains art by C Bangs).
Matloff, G. L., 'The Zero-Point Energy (ZPE) Laser and Interstellar Travel', in *NASA Breakthrough Propulsion Physics Workshop Proceedings, NASA/CP-1999-208694*, ed. M. Millis, NASA Lewis (Glenn) Research Center, August 12–14, 1997.

Matloff, G. L. and Johnson, L., 'Magnetic Surfing: Reformulation of Lenz's Law and Applications to Space Propulsion', AIAA-2000-3338.

Marx, G., 'Mechanical Efficiency of Interstellar Vehicles', *Acta Astronautica*, **9**, 131–139 (1963).

Miley, G., 'Possible Evidence of Anomalous Energy Effects in H/D-Loaded Solids – Low Energy Nuclear Reactions (LENRs)', in *NASA Breakthrough Propulsion Physics Workshop Proceedings, NASA/CP-1999-208694*, ed. M. Millis, NASA Lewis (Glenn) Research Center, August 12–14, 1997.

Millis, M., 'NASA Breakthrough Propulsion Physics Program', in *Missions to the Outer Solar System and Beyond, 2nd IAA Symposium on Realistic Near-Term Scientific Space Missions*, ed. G. Genta, Levrotto & Bella, Turin, Italy (1998), pp. 103–110.

Minami, Y., 'Possibility of Space Drive Propulsion', IAA-94-IAA.4.1.658.

Moeckel, W. E., 'Thrust-Sheet Propulsion Concept using Fissionable Elements', *AIAA Journal*, **15**, 467–475 (1977).

Mourou, G. A., Barty, C. P. J. and Perry, M. D., 'Ultrahigh-Intensity Lasers: Physics of the Extreme on a Tabletop', *Physics Today*, **51**, 22–28 (January 1998).

Morgan, J. A., 'Neutrino Propulsion for Interstellar Spacecraft', *Journal of the British Interplanetary Society*, **52**, 424–428 (1999).

Noever, D. and Koczor, R., 'Granular Superconductors and Gravity', in *NASA Breakthrough Propulsion Physics Workshop Proceedings, NASA/CP-1999-208694*, ed. M. Millis, NASA Lewis (Glenn) Research Center, August 12–14, 1997.

Ohanian, H. C., *Physics*, 2nd edn, Norton, New York (1989), Chapter 41.

Oliver, B. M., 'A Review of Interstellar Rocketry Fundamentals', *Journal of the British Interplanetary Society*, **43**, 259–264 (1990).

Perry, M., 'Crossing the Petawatt Threshold', *Science and Technology Review*, 4–11 (December, 1996).

Perry, M. D. and Mourou, G., 'Terawatt to Petawatt Subpicosecond Lasers', *Science*, **264**, 917–924 (1994).

Puthoff, H. E., 'Can the Vacuum be Engineered for spaceflight applications?: Overview of Theory and Experiments', in *NASA Breakthrough Propulsion Physics Workshop Proceedings, NASA/CP-1999-208694*, ed. M. Millis, NASA Lewis (Glenn) Research Center, August 12–14, 1997.

Sears, F. W., Zemansky, M. W. and Young, H. D., *University Physics, Part 1*, 5th edn. Addison–Wesley, Reading, MA (1977), Chapter 14.

Seward, C., 'Propulsion and Energy Generation using the Electron Spiral Toroid', in *NASA Breakthrough Propulsion Physics Workshop Proceedings, NASA/CP-1999-208694*, ed. M. Millis, NASA Lewis (Glenn) Research Center, August 12–14, 1997.

Shepherd, L. R., 'Interstellar Flight', in *Realities of Space Travel*, ed. L. J. Carter, Putnam, London (1958).

Szames, A., *The Biefeld–Brown Effect*, (in French and English), ASZ editions, Boulogne, France (1998).

Winglee, R., Slough, J., Ziemba, T. and Goodson, A., 'Mini-Magnetosphere Plasma Propulsion (M2P2): High Speed Propulsion Sailing the Solar Wind', presented at STAIF 2000 Conference, University of New Mexico, Albuquerque, NM, January 30–February 3, 2000.

Yilmaz, H., 'The New Theory of Gravitation and the 5th Test', in *NASA Breakthrough Propulsion Physics Workshop Proceedings, NASA/CP-1999-208694*, ed. M. Millis, NASA Lewis (Glenn) Research Center, August 12–14, 1997.

Zhang, S. N., Cui, W. and Chen, W., 'Black Hole Spin in X-Ray Binaries: Observational Consequences', *Astrophysical Journal*, **482**, L155–L158 (1997).

Zhang, S. N., Yu, W. and Zhang, W., 'Spectral State Transitions in Aquila X-1: Evidence for 'Propeller Effects', *Astrophysical Journal*, **494**, L71–L74 (1998).

Also see papers in a special issue of *Journal of the British Interplanetary Society*, **52**, No. 9 (1999).

10

Of stars, planets and life

And there is a star in the southern sky,
the most magnificent star that I have ever seen,
and I am beginning to know its name,
Alpha Centauri

Robert Ardrey, *African Genesis* (1961)

No matter how we travel to the stars, we must learn all we can about the environments surrounding these distant suns. Only then can we dispatch out robot proxies or begin outfitting the ships to be occupied by humanity's first interstellar pioneers.

If Earth-like planets are rare in the Milky Way Galaxy, interstellar expeditions will also be infrequent. Perhaps only a civilization threatened by the ultimate catastrophe of its star's demise would then attempt the colonisation of a neighbouring star's comet cloud.

Because of their small mass relative to the parent body and the great distances involved, we cannot yet detect Earth-like worlds orbiting even the nearest stars. But the recent discoveries of many Jupiter-sized worlds orbiting nearby stars encourages those astronomers who dream of constructing a new generation of telescopes capable of imaging these suspected, tantalising blue planets.

10.1 A SHORT HISTORY OF EXTRASOLAR PLANET DETECTION EFFORTS

For several decades, astronomers have attempted to detect Jupiter-sized planets orbiting nearby stars by observing the wobble in a star's image produced by hypothetical invisible objects less than 1% of the star's mass and orbiting that star. To obtain reasonably promising results using this astrometric technique, it is necessary to concentrate upon very near low-mass, high-proper motion stars. Hundreds or thousands of photographic images of the near star's position relative to more distant background stars are necessary for the astrometric technique to work (van de Kamp, 1967).

The work is painstaking and the computational requirements are exhaustive. Many non-planetary factors can mask or imitate the stellar wobble, which amounts to about 1 arcsec per year. It is therefore perhaps not surprising that the early announcements of planet detections turned out to be false alarms. The classic false alarm is the case of Barnard's Star, a metal-poor, high-proper motion red dwarf that is the second nearest star system to the Sun. As reviewed by S. J. Dick, Peter van de Kamp and Sarah Lippincott, of Sproul Observatory at Swarthmore University, analysed more than 2,000 photographic images of this star taken using the Sproul refractor between 1916 and 1961. They concluded that one or two objects, approximately of Jupiter's mass, orbited this star. Although the astronomical and astronautical communities were quite excited about the announced discovery, it was of course necessary to confirm it using photographic plates exposed at other observatories. This was attempted by George Gatewood of Allegheny Observatory, and Heinrich Eichorn of the University of Southern Florida, who evaluated 241 plates of Barnard's Star taken at two observatories between 1916 and 1971. The proper-motion wobble was not confirmed, and most astrometric astronomers now believe that the spurious Barnard's Star planets were due to a change of the cell of the object glass of the Sproul telescope in 1949.

Even though astrometry has failed to yield the first confirmed extrasolar-planet discovery, astronomers had better luck with other approaches during the 1980s and 1990s. As reviewed by Dick, one approach that yielded fruit was spectroscopic planet detection, which was pioneered by Bruce Campbell of the University of British Columbia during the 1980s.

To detect an extrasolar planet spectroscopically, it is necessary to observe the spectrum of the planet's primary star as the planet and star revolve around their common centre of mass. The Doppler shift of the spectral lines caused by a large planet's gravitational tug upon its more massive primary can be detected with sufficiently sensitive equipment. Stellar radial velocity changes of about $10\,\mathrm{m\,s}^{-1}$ are currently observable using this approach.

Another significant discovery of the 1980s, as reviewed by Ken Croswell, was the infrared space telescope imaging of preplanetary circumstellar dust rings encircling some young nearby stars. Nearby main sequence stars possessing these dust rings include β Pictoris, ε Eridani, Fomalhaut and Vega.

But the first confirmed discoveries of extrasolar planets during the early 1990s did not accompany such well-behaved, stable stars. As reviewed by D. E. Fisher and M. J. Fisher, these strange planets orbit the pulsars resulting from supernova explosions, and were detected by planet-caused variations in the radio emissions from the central pulsars. Even though life-bearing worlds are very unlikely in the extreme environments of such stellar graveyards, these discoveries indicated that extrasolar planets must indeed be common within our Galaxy.

Finally, radial-velocity measurements of nearby Sun-like stars began to bear fruit in the mid-1990s. Two planet-searching teams – one in Switzerland, headed by Michel Mayor of Geneva Observatory (Mayor *et al.*, 1997), and one at Lick

Observatory, directed by Geoff Marcy of San Francisco State University (Butler and Marcy, 1997) – began to detect the first of dozens of large extrasolar planets. Some of these worlds reside in solar systems more or less like our own; others orbit their parent stars in highly elliptical paths; and still others are an unexpected class of 'hot Jupiters' – giant planets orbiting ten million kilometres or so from the primary star. The two groups have been able to confirm each other's discoveries. Most significantly – as described by Alan Boss in 1996 – the Hubble Space Telescope was able to photograph a brown dwarf (an object intermediate in size between a jovian planet and a tiny star) that had been detected orbiting the nearby red dwarf Gliese 229. This photograph was the first independent confirmation of the radial velocity method of planet detection.

Another validation of this approach has appeared very recently. As described by John Nobel Wilford in the 16 November 1999 edition of *The New York Times*, and by Govert Schilling in the 19 November 1999 issue of *Science*, the shadow of a giant planet, whose orbit around HD 209458 (a Sun-like star 153 light years distant in the constellation Pegasus) passes directly between the parent star and our Sun every 3.5 days, was observed photometrically, as predicted by radial motions of that star.

We can detect large Jupiters using radial-velocity measurements (and possibly astrometry), but cannot yet image them orbiting normal stars. Brown dwarfs have been imaged, as has a possible giant 'rogue' planet apparently being ejected from a binary star system (Schilling, 1999). It is only a matter of time until the next generation of space and terrestrial telescopes succeeds in imaging an extrasolar planet of jovian or even terrestrial size.

10.2 METHODS OF IMAGING EXTRASOLAR PLANETS

The technical problems of imaging extrasolar planets are even greater than those of remotely viewing NEOs (discussed in Chapter 1). Not only is the planet very faint and very far away from us, but is in proximity to an object that is approximately equivalent to our Sun in respect to radiant emissions.

There are three basic approaches to imaging an extrasolar planet or brown dwarf orbiting a main-sequence star. All of them attempt to either reduce received stellar radiant emissions or enhance received planetary radiant emissions. These approaches are to use a very large telescope aperture, to use a smaller telescope with an occulting 'mask' to cover the central image of the object's primary star, and to select spectral regions in which the planet's radiant emissions are enhanced. These strategies, of course, are not mutually exclusive.

One of the first suggestions was that of Gerard K. O'Neill, who in 1968 proposed the construction of an orbital array of mirrors with separation of more than 100 m and an effective resolving power about 25× that of the 5-m reflecting telescope on Mount Palomar in Southern California. As shown in Figure 10.1, all the elements of such a high-resolution orbital interferometer would have a common focus. Although the device would have much less light-gathering power than a solid mirror 100 m in diameter, it would be immensely easier to construct and maintain. As discussed

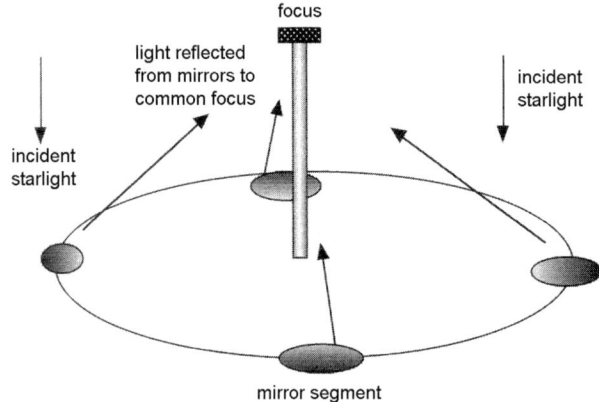

Fig. 10.1. An optical interferometer, consisting of many widely separated mirrors with a common focus.

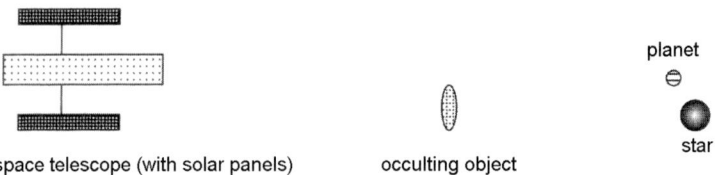

Fig. 10.2. The use of an occulting disk for enhancing the observability of an extrasolar planet.

below, the problems of resolving planets orbiting other stars were underestimated by O'Neill and other early authors.

Even a smaller space telescope could be utilised as an extrasolar planet imager if an occulting disc of some sort were used to partially mask the light from the primary star (Figure 10.2). This idea was first investigated for application to the Hubble Space Telescope (HST) by NASA astronomer Nancy Roman in 1959, and was later elaborated in papers by Fennelly *et al.* and Elliot. If the early assumptions on planet–star contrast had been correct, the HST could detect at least Jupiter-sized planets orbiting nearby stars by using self-propelled occulters, occulting filters in the telescope's optical path, or even the Moon's limb. Such an occulting device reduces light received by imaging equipment from the primary star by blocking the central portion of the star's image. Since the planet will be observed in outer fringes of the star's Airy diffraction pattern, advanced image-processing techniques are required to view the planet, even with stellar occultation.

A pivotal paper reviewing various approaches to extrasolar-planet imaging was published by Jill Tarter and colleagues in 1986. When imaging a planet orbiting in the proximity of a distant star, a significant factor is B_{ep}, the brightness ratio of planet to starlight in the received image. This can be approximated in the limits of short observing wavelength (sw) and long observing wavelength (lw) using the equations

$$B_{ep,sw} \sim \left(\frac{2R_{planet}}{D_{planet-star}}\right)^2, \quad \left(\frac{hc}{KT_{star}} > \lambda\right) \quad (10.1)$$

$$B_{ep,lw} \sim \frac{T_{eff,planet} R_{planet}^2}{T_{star} R_{star}^2}, \quad \left(\frac{hc}{KT_{eff,planet}} < \lambda\right) \quad (10.2)$$

where R_{planet} and R_{star} are respectively planet and star radii, T_{star} is the star's surface (or effective black-body) temperature in degrees K, T_{eff}, planet is the planet's effective or black-body temperature in degrees K, $D_{planet,star}$ is the planet–star separation, c is the speed of light, k is Boltzmann's constant (1.38×10^{-23} Joule/degrees K), h is Planck's constant (6.63×10^{-34} Joule-seconds), and λ is the observing wavelength in metres. For the case of our Sun and Earth with, respectively, effective temperatures of 6,000 K and 300 K, the shortwave (sw) approximation is accurate for wavelengths shorter than about 2.4 µm, and the longwave (lw) approximation is accurate for wavelengths longer than about 50 µm.

Exercise 10.1 Evaluate the ranges of applicability for equations (10.1) and (10.2) for the case of Jupiter-like planets ($T_{eff,planet} = 125$ K) orbiting stars with surface temperatures of 4,000, 6,000 and 7,000 K.

As discussed by both Tarter *et al.* and Black (1980), the correct version of Rayleigh's criterion applied to planetary detection yields an expression for the angular separation between a planet and its much brighter primary star that is detectable:

$$\Delta\theta = \frac{3.56 \times 10^4 \lambda}{B_{ep}^{1/3} R_{tele}} \text{ arcsec} \quad (10.3)$$

where R_{tele} is the telescope's effective aperture in metres.

Exercise 10.2 Assume that you are observing an extrasolar planet with a 50-m radius space telescope at a wavelength of 1 µm. First estimate the planet's brightness ratio relative to its primary star using equation (10.1) if the planet's radius is 6,000 km and it is 1 AU from its star. Then use equation (10.3) to calculate the detectable separation of the planet from its primary star. At a distance of 1 parsec (3.26 light years) from the Sun, the Earth will have an angular separation from the Sun of 1 arcsec. What is the maximum distance at which this telescope could separate an Earth-like planet from its Sun-like primary?

One possible approach to extrasolar-planet imaging in the near future has been reviewed by Fisher *et al.* (1997). A number of high-altitude terrestrial interferometers with baselines sufficient to resolve nearby extrasolar planets in the infrared have begun operation, or will do so in the near future. Using speckle interferometry (in which observations of a laser reflected from the Earth's upper atmospheric layers are used to correct for atmospheric turbulence) and adaptive optics (in which the telescope corrects internally for variations in 'seeing'), jovian or even Earth-like planets orbiting the nearest stars may soon be imaged using these telescopes. Applications of adaptive optics have been reviewed by Thompson (1995). McAlister (1977) has described the use of speckle interferometry in planet detection.

In 1997, Richard Terrile and Christ Ftaclas suggested that a 1.5-m infrared (IR) telescope mounted in a balloon-borne observatory and equipped with appropriate occulting equipment could image Jupiter-like worlds orbiting nearby stars. The advantage of an infrared telescope for this purpose is that Sun-like stars radiate only a small fraction of their energy in the infrared range, while planets reradiate most of their absorbed solar energy in the infrared range of the electromagnetic spectrum.

In 1996, Jeff Van Cleve and his colleagues suggested that even a very modest infrared space telescope could map diffuse emissions from Solar System objects and serve as a testbed for more elaborate infrared space telescopes to actually image Earth-like and Jupiter-like planets orbiting nearby stars. But as described by Donald Goldsmith in 1996, the ultimate near-term infrared extrasolar-planet imager might be a spacecraft-mounted interferometer with mirror separations as great as 1 km, located 3–4 AU from the Sun. The challenge of maintaining nanometer-tolerance mirror alignment remotely over such distances is not to be minimised!

The size of infrared space interferometers dedicated to extrasolar-planet imaging can be reduced with application of appropriate occulting devices. Self-propelled occulter design has been discussed in the literature by a number of astronomers, including Marchal (1984) and Jordan *et al.* (1999), who have proposed applying this technology to the Next Generation Space Telescope (NGST).

10.3 EXTRASOLAR PLANETS FOUND TO DATE THAT ORBIT SUN-LIKE STARS

It is impossible to provide an up-to-date survey of all planets that have been discovered orbiting nearby stars, and the reader should therefore keep in mind that this section is based upon planet-detection data available in November 1999. Without doubt, many new worlds will be added to the list by the time this book is published.

The discovered extrasolar planets tabulated here are the latest confirmed planets (with one noted exception) as of 20 November 1999. These are all from *The Encyclopedia of Extrasolar Planets*, a website (http://www.obspm.fr/planets) maintained by Jean Schneider of Observatoire de Paris.

Table 10.1 presents the more-or-less 'normal' solar systems that have been discovered and confirmed accompanying nearby Sun-like stars. At least some of the planets orbiting a star should be in orbits approximating planets in our Solar System, and planetary orbital eccentricities are less than 0.1 to qualify for this list. The significance of stellar spectral classes is discussed in the next section of this chapter.

One detected but currently unconfirmed nearby solar system that may eventually rate inclusion in Table 10.1 is the 1996 astrometric discovery by George Gatewood of one or more jovians orbiting one of the nearest stars, Lalande 21185, at 8.5 light

Table 10.1. Confirmed 'normal extrasolar solar systems' (November 1999)

Star name	Star spectral class	Sun–star separation	Minimum planet mass (in Jupiter masses)	Star–planet separation	Planet orbital period (days)
υ And	F8V	44 light years	0.71	0.059 AU	4.62
			2.11	0.83	241
			4.61	2.50	1,267
47 UMa	G1V	46	2.41	2.10	1,100
HD 222582	G5	137	5.4	1.35	576

Table 10.2. Confirmed 'close Jupiters' (November 1999)

Star name	Star spectral class	Sun–star separation	Minimum planet mass (in Jupiter masses, M_j)	Star–planet separation	Planet orbital period (days)
HD 75289	G0V	94 light years	0.42	0.046 AU	3.51
51 Peg	G2V	50	0.47	0.05	4.3
HD 187123	G5	163	0.52	0.042	3.1
HD 209458	G0V	153	0.63	0.045	3.5
HD 192263	K2V	65	0.76	0.15	23.9
55 Cnc	G8V	41	0.84	0.11	14.7
HD 130322	K0III	98	1.08	0.088	10.7
ρ CrB	G0V	57	1.1	0.23	39.7
HD 195019	G3IV-V	122	3.43	0.14	18.3
Gl86	K1V	36	4	0.11	15.8
τ Boö	F6IV	51	3.87	0.046	3.31

years from the Sun. If a 90% Jupiter mass planet does indeed orbit this star every 5.8 years, such a nearby solar system may provide an interesting target for an early interstellar exploration mission.

Known and confirmed 'close Jupiters' are presented in Table 10.2. Each of these planets is closer than 0.3 AU to its primary star, and has an orbital eccentricity of less than 0.1.

A final class of confirmed extrasolar planets is the 'eccentric Jupiters', which have an orbital eccentricity greater than 0.1. Confirmed members of this class are listed in Table 10.3.

Of the two dozen or so confirmed extrasolar planets or solar systems, only about 10% are in paths roughly resembling those of the Sun's planets. The most prevalent planet type in the sample is the eccentric Jupiters. But because only the minimum planet masses are well known at present, some of the eccentric Jupiters may turn out to be brown dwarfs rather than giant planets.

Table 10.3. Confirmed 'eccentric Jupiters' (November 1999)

Star name	Star spectral class	Sun–star separation	Minimum planet mass	Mean Star–planet separation	Planet orbital period (days)	Eccentricity
HD 37124	G4V	10 light years	1.04 M_j	0.0585 AU	155	0.19
HD 177830	K0	192	1.28	1.0	391	0.43
HD 217107	G8IV	64	1.28	0.07	7.1	0.14
HD 210277	G0V	69	1.28	1.10	437	0.45
HD 134987	G5V	82	1.58	0.78	260	0.25
Gliese 876	M4V	15	2.1	0.21	60.9	0.27
HR 810	G0V	~ 50	2.26	0.93	320	0.16
14 Her	K0V	59	3.3	2.5	1619	0.35
HD 168443	G5	123	5.04	0.28	57.9	0.54
HD 10697	G5IV	98	6.59	2.0	1083	0.12
70 Vir	G4V	59	6.6	0.43	116.7	0.4
HD 114762	F9V	132	11	0.3	84	0.33

It is not know whether the statistical breakdown of planet types in the sample is typical for the Universe or a function of the radial-velocity planet-detection techniques. Another mystery is the large number of close Jupiters in the sample. Perhaps, as speculated by Murray *et al.* (1998), two giant planets form close to each other under certain conditions. One might be expelled from the infant solar system or flung into an eccentric orbit, and its companion might be pushed by gravitational interaction with its giant companion close to the parent star. As discussed by Eric Sandquist in 1999, perhaps some stars actually 'swallow' very close Jupiters under certain conditions.

But perhaps the most exciting mystery about our stellar neighbours is the probability of life-bearing worlds. Planets are certainly there, perhaps orbiting most of the stars in the cosmos; but how many of these are worlds on which we could live, and how many might have evolved higher life of their own?

10.4 HOW COMMON ARE LIFE-BEARING WORLDS?

To estimate the relative frequency of life-bearing worlds in the galaxy, a good starting point is the Hertzsprung–Russell (H–R) diagram of stellar classification and evolution (Figure 10.3). Stellar spectral classes are further subdivided into subclasses (for example, the hottest G stars are G0, and the coolest are G9). The Sun is currently a G2V main sequence dwarf. It has been on the main sequence for about 5×10^9 years, and will depart the main sequence about 5×10^9 years in the future, expanding first to become a Class IV subgiant (intermediate in the H–R diagram between luminosity classes III and V) and continue expanding until it becomes a giant.

Sec. 10.4] How common are life-bearing worlds? 141

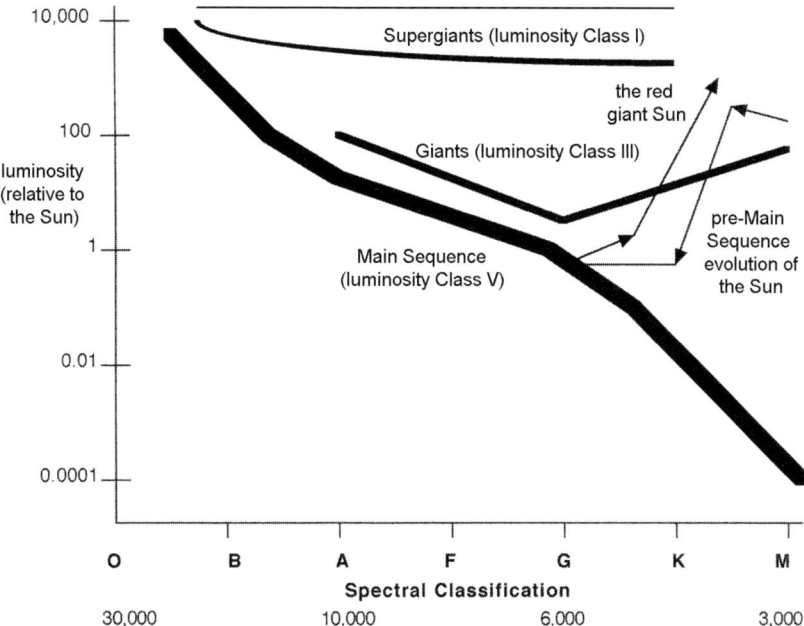

Fig. 10.3. The Hertzsprung–Russell diagram, showing spectral classes, surface temperatures, luminosity classes and solar evolution. (Illustration derived from E. Chaison and S. McMillan, *Astronomy*, second edition, Prentice Hall, 1998.)

Giants are not good locales to search for extrasolar life-bearing worlds, partly because a star remains in this phase for 'only' about 10^8 years. After departing the giant phase in the distant future, our Sun will have used most of its thermonuclear fuel reserves. It will contract and decrease in luminosity, cross the main sequence again, and end its career as a subluminous white dwarf star, in the lower left corner of the H–R diagram.

More than half the stars in the sky are binaries. As reviewed in Chapter 2, some binaries such as our near-neighbour α Centauri may have life-bearing worlds orbiting either of the binary members.

Hot, massive O stars have a main-sequence life of only a few million years, but cool, low-mass M stars reside on the main sequence for as long as 10^{12} years. Astronomers have long suspected that F, G and K stars have a sufficiently long main-sequence lifetime (billions of years) for life-bearing worlds to evolve. However, habitable planets orbiting M stars were considered unlikely because of the proximity of such planets to the cool parent star. But recent computer models of planetary habitability reported by Kasting *et al.* (referenced in Chapter 2) and Doyle *et al.* indicates that habitable worlds orbiting M-type main sequence dwarfs are not impossible.

As discussed in Chapter 2, the dimensions of the ecosphere – the lifezone – orbiting main-sequence stars may have been seriously underestimated. If non-stellar heat sources affect planets or satellites far from a star – as may be the case for

Table 10.4. The nearest nearby stars that might possess habitable planets

Name	Spectral class	Distance	Habitable planet probability
α Centauri A	G2V	4.3 light years	0.054
α Centauri B	K6V	4.3	0.057
ε Eridani*	K2V	10.7	0.033
τ Ceti	G8V	11.9	0.036
70 Ophiuchi A	K1V	16.7	0.057
η Cassiopeiae A	G0V	19.2	0.057
σ Draconis	K0V	18.5	0.036
36 Ophiuchi A	K2V	17.7	0.023
36 Ophiuchi B	K1V	17.7	0.020
HR 7703 A	K3V	18.4	0.020
δ Pavonis	G6V	18.8	0.057
82 Eridani	G5V	20.3	0.057
β Hydri**	G2IV	20.5	0.037
HR 8832***	K3V	21.4	0.011

All probabilities are from S. H. Dole, *Planets for Man* (1964). Except where noted, all spectral types and stellar distances are from E. G. Mallove and G. L. Matloff, *The Starflight Handbook* (1989).
* Habitable planet probability may be much lower, as the star is very young.
** Planets may be uninhabitable, as the star has left the main sequence. Star spectral class from Dravins *et al.*
*** Star spectral class and distance from Dole.

Jupiter's satellite Europa – those worlds could support life. The calculations of habitable-planet probabilities for nearby stars published by Stephen Dole in 1964 (see Chapter 2 bibliography) may therefore be greatly underestimated. Doles' results are reproduced in Table 10.4. Although his probability estimates are generally probably too low, two nearby stars should be removed from his list. As discussed earlier in this chapter, a pre-planetary disc has been observed around ε Eridani, which indicates that this star is probably too young for habitable planets to have evolved. As discussed by Dravins *et al.* in 1998, β Hydri has apparently ended its main-sequence career and should be reclassified as a subgiant. Habitable planets are not likely to orbit such an aged, expanding star.

10.5 TERRAFORMING: A WAY TO INCREASE THE ODDS

While the likelihood of finding a life-bearing world orbiting a given star may be somewhat greater than Dole's habitable planet probabilities listed in Table 10.4, it is by no means assured that humans could survive without spacesuits on all life-bearing worlds. It is true that a human colony could survive on Mars and perhaps on Europa, but it would be necessary for the colonists to live underground or under a geodesic dome and be supported by a closed ecological system.

To improve the probability of an interstellar colony succeeding, we should learn the art of 'terraforming' – altering a planet's environment to suit our tastes. In science fiction, terraforming is fast and easy; but as discussed by Martyn Fogg in his 1995 review of the subject, to render a Mars-like planet comfortable for human colonists will be a centuries-long project.

Some proposed terraforming techniques use brute force. Colonists may direct comets to bombard a dry, Mars-like world to produce ample oceans; other techniques are more subtle – huge, space-based solar sails may be used as reflectors or sunshades to alter the amount of insolation received by the planet. Biological techniques include genetically altered organisms to convert a thin CO_2 atmosphere to oxygen using photosynthesis.

Because terraforming will be time-consuming and demand a large local industrial and biological base, star travellers may elect to travel in huge, slow worldships rather than in small, fast starships. During the centuries or millennia that are apparently required to achieve successful terraforming, the colonists might choose to continue living in the large, comfortable and well-equipped ship that carried their ancestors from Earth.

But there is an ethical question to be dealt with here. Is it appropriate for humans to modify an environment to which alien, primitive life has adapted?

10.6 BIBLIOGRAPHY

Black, D. C., 'In Search of other Planetary Systems', *Space Science Review*, **25**, 35–81 (1980).

Boss, A. P., 'Extrasolar Planets', *Physics Today*, **49**, No. 9, 32–38 (September 1996).

Butler, R. P. and Marcy, G. W., 'The Lick Observatory Planet Search', in *Astronomical and Biochemical Origins and the Search for Life in the Universe*, ed. C. B. Cosmovici, S. Bowyer and D. Werthimer, Editrice Compositori, Bologna, Italy (1997), pp. 331–342.

Croswell, K., *Planet Quest*, The Free Press, New York (1997).

Dick, S. J., *The Biological Universe*, Cambridge University Press, New York (1986).

Doyle, L., Billingham, J. and Devincenzi, D., 'Circumstellar Habitable Zones: An Overview', IAA-95-IAA.9.1.03.

Dravins, D., Lindgren, L. and VandenBerg, D. A., 'Beta Hydri (G2IV): A Revised Age Estimate for the Closest Subgiant', *Astronomy and Astrophysics*, **330**, 1077–1079 (1998).

Elliot, J. E., 'Direct Imaging of Extra-Solar Planets with Stationary occultations Viewed by a Space telescope', *Icarus*, **35**, 156–163 (1978).

Fennelly, A. J., Matloff, G. L. and Frye, G., 'Photometric Detection of Extrasolar Planets using L.S.T.-Type Telescopes', *Journal of the British Interplanetary Society*, **28**, 399–404 (1975).

Fisher, D. E. and Fisher, M. J., *Strangers in the Night*, Counterpoint, Washington, DC (1998).

Fisher, R. J., Houstan Jr., J. B. and Stuhlinger, T. W., 'Infrared Interferometers for Observing Extrasolar Planets', in *Astronomical and Biochemical Origins and the Search for Life in the Universe*, ed. C. B. Cosmovici, S. Bowyer and D. Werthimer, Editrice Compositori, Bologna, Italy (1997), pp. 367–373.

Fogg, M. J., *Terraforming: Engineering Planetary Environments*, Society of Automotive Engineers, Warrendale, PA (1995).

Gatewood, G., 'Possible Planet in Lalande 21185', *Bulletin American Astronomical Society*, **28**, 885 (1996).

Goldsmith, D., 'Seeking out Strange New Worlds', *Science*, **271**, 588 (1978).

Jordan, J. E., Schultz, A. B., Schroeder, D., Hart, H. M., Bruhweiler, F., Fraquelli, D., Hamilton, F., DiSanti, M., Melodi, R., Cheng, K.-P., Miskey, C., Kochte, M., Johnson, B. and Sami, F. M., 'Enhancing NGST Science: Umbras', in *Proceedings NGST Science and Technology Exposition*, Hyannis, MA, September 13–16, 1999 (Astronomical Society of the Pacific).

Marchal, C., 'Concept of a Space Telescope Able to See the Planets and Even the Satellites Around the Nearest Stars', *Acta Astronautica*, **12**, 195–201 (1985).

Mayor, M., Queloz, D., Udry, S. and Halbwachs, J. L., 'From Brown Dwarfs to Planets', in *Astronomical and Biochemical Origins and the Search for Life in the Universe*, ed. C. B. Cosmovici, S. Bowyer and D. Werthimer, Editrice Compositori, Bologna, Italy (1997), pp. 313–330.

McAlister, H. A., 'Speckle Interferometry as a Method of Detecting Nearby Extrasolar Planets', *Icarus*, **30**, 789–792 (1977).

Murray, N., Hansen, B., Holman, M. and Tremaine, S., 'Migrating Planets', *Science*, **279**, 69–72 (1998).

O'Neill, G. K., 'A High-Resolution Orbiting Telescope', *Science*, **160**, 843–847 (1968).

Roman, N. G., 'Planets of Other Stars', *Astronomical Journal*, **64**, 344–345 (1959).

Schilling, G., 'Stellar Small Fry or Runaway Planet', *Science*, **285**, 1471–1472 (1999).

Schilling, G., 'Shadow of an Exoplanet Detected', *Science*, **286**, 1451–1452 (1999).

Sandquist, E., 'More than Just a Planet, it's a Meal', *Mercury*, **28**, No. 1, 10–13. (January/February, 1999).

Tarter, J. C., Black, D. C. and Billingham, J., 'Review of Methodology and Technology Available for the Detection of Extrasolar Planetary Systems', *Journal of the British Interplanetary Society*, **39**, 418–424 (1986).

Terrile, R. J. and Ftaclas, C., 'Direct Detection of Extrasolar Planetary Systems from Balloon Borne Telescopes', in *Astronomical and Biochemical Origins and the Search for Life in the Universe*, ed. C. B. Cosmovici, S. Bowyer and D. Werthimer, Editrice Compositori, Bologna, Italy (1997), pp. 359–366.

Thompson, L. A., 'Adaptive Optics in Astronomy', *Physics Today*, **47**, No.12, 24–31 (December 1994).

Van Cleve, J., 'Infrared Inferences of Planetary Systems Among the Nearby Stars', *Journal of the British Interplanetary Society*, **49**, 3–6 (1996).

van de Kamp, P., *Principles of Astrometry*, Freeman, San Francisco CA (1967).

11

Life between the stars

To seek it with thimbles,
To seek it with care;
To pursue it with forks and hopes;
To threaten its life with a railway share;
To charm it with smiles and soap!

          Lewis Carroll, *The Hunting of the Snark* (1891)

As we have seen in the previous chapters, there are many roads to the stars and a wide variety of planetary environments await the first pioneers from Earth. But unless some form of breakthrough propulsion proves possible and practical, the duration of even the shortest interstellar journies will be decades or centuries, and the planets awaiting the first pioneers may require centuries of terraforming, since exact duplicates of the Earth may not be common.

There are environmental hazards to be considered in planning interstellar missions by humans: principally production of artificial gravity and protection from impacts by interstellar dust grains and cosmic rays. As discussed below, these can be addressed technologically.

Other technological issues revolve around the maintenance of a human community's life-style on an interstellar voyage: what are the energy options to support life between the stars, and can we design appropriate closed ecological systems? These problems also have technological solutions.

Perhaps the most intractable problems to be considered by interstellar mission designers are sociological. How do we engineer the small population of a starship so that the crew remains sane and effective during the long decades in close proximity to each other, and in grand isolation from everyone else? We might approach this by constructing a 'worldship' – a starship large enough to include a very diverse community in a simulated planetary environment. Alternatively, we might instead sleep our way to the stars.

11.1 ENVIRONMENTAL OBSTACLES TO INTERSTELLAR FLIGHT, AND THEIR REMOVAL

One potential barrier to interstellar travel is interstellar dust. As a starship moves through the interstellar medium at speeds of 0.005 c or higher, even tiny dust grains have a large kinetic energy relative to the ship.

Collisions with interstellar dust grains

As reviewed by Chaisson and MacMillan's *Astronomy Today* (listed in Chapter 2 bibliography), recent studies of the interstellar medium have demonstrated that typical interstellar dust grains have dimensions of 0.1–1 µm, have average interstellar-medium densities of about 1,000 dust grains per cubic kilometre (less in the rarefied interstellar medium in the Solar System's vicinity), are elongated in shape, and are affected by the interstellar magnetic field. Infrared evidence indicates that water, methane and ammonia ices are constituients of the interstellar dust grains, as are graphites, silicates and iron.

Consider an interstellar dust grain with a size of 0.2 µm (2×10^{-7} m). Approximating this grain by a sphere, its volume is about 4×10^{-21} m^3. If the specific gravity of the grain is that of graphite (about 2), its mass (M_{dust}) is about 10^{-17} kg. The number of dust impacts on the ship per second per square metre is equal to $3 \times 10^8 \rho_{dust} \beta_s$, where ρ_{dust} is the dust grain density per cubic metre, and β_s is the spacecraft velocity as a fraction of the speed of light (relative to the dust grain). Assuming that $\rho_{dust} = 10^{-6}$ m^{-3}, there are about 1.5 impacts s^{-1} m^{-2} when $\beta_s = 0.005$ c, and 30 impacts s^{-1} m^{-2} when $\beta_s = 0.1$ c.

In a totally inelastic collision, the kinetic energy transferred to the ship from the dust grain is $0.5 M_{dust} (3 \times 10^8 \beta_s)^2$ J. For $M_{dust} = 10^{-17}$ kg, the kinetic energy transferred to the ship by each totally inelastic collision with a dust grain is about $0.5(\beta_s)^2$ J. At $\beta_s = 0.005$ c, each dust-grain collision transfers a maximum of 1.25×10^{-5} J to the ship. At $\beta_s = 0.1$ c, each collision transfers a maximum of 0.005 J to the ship.

Applying the above discussion of the number of impacts against the starship's structure, about 2×10^{-5} W m^{-2} of radiant power must be radiated from the starship's structure at 0.005 c, and about 0.15 W m^2 of radiant power must be radiated from the starship's structure at 0.1 c. Clearly, the heating effect of interstellar dust grain collisions will be negligible.

However, as reviewed in Mallove and Matloff's *The Starflight Handbook*, starship erosion by collisions with interstellar dust grains might be a hazard. Early work in this field has also been reviewed by Tony Martin, in a contribution to the *Project Daedalus Final Report*.

There are a number of alternatives to prevent or reduce starship erosion by dust grains. A low-mass wire mesh or aerosol-cloud dust-guard could be placed in front of a cruising starship to break the grains apart. This might not be necessary at vehicle speeds less than about 0.01 c, since the dust-grain kinetic energy is so low.

Martin reviewed the many contending theories of impacting dust-grain interaction with starship structures at higher starship velocities. The major uncertainty is the fraction of dust-grain kinetic energy that is transferred to the spacecraft during an impact. Conservatively, Martin assumed that the dust guard required to protect a large starship moving at 0.15 c would be in the vicinity of 50,000 kg – not a huge mass increment.

As reviewed in *The Starflight Handbook*, there is uncertainty about dust-grain impact effects upon laser light sails. If the impacting grain passes through the sail without depositing too much of its kinetic energy, damage will be minimal.

Exercise 11.1 Although ordinary grains of interstellar dust may pose a minor hazard, there might be occasional larger dust particles up to 0.1 kg in mass. Calculate the kinetic energy relative to the ship of large dust particles massing 10 g and 100 g at starship velocities of 0.005, 0.02, and 0.1 c. Some form of active defense might be necessary for protection against large grain impacts. They could be detected by radar and deflected or destroyed using particle beams or lasers.

Because interstellar dust grains contain some iron and are affected by magnetic fields, the ramjet fields, magsails and mini-magnetospheres described in previous chapters might deflect them. If the grains are sufficiently tenuous, the rapidly varying ship-generated magnetic field might dissipate the grains long before they encounter the starship.

Cosmic radiation and crew health

Much of the pre-1992 work on cosmic ray effects and protection is surveyed by John Mauldin in *Prospects for Interstellar Travel* (referenced in Chapter 4). There are two basic sources for these high-energy, ionised, subatomic particles: the Sun and the Galaxy. Galactic cosmic rays are partially shielded from the inner Solar System by the interplanetary magnetic field. More significant for the survival of terrestrial life is the shielding effects of Earth's magnetosphere and atmosphere.

Recognition of possible health consequences from a long-term exposure to Galactic cosmic rays, especially those more massive than α particles ('high-Z' cosmic rays) occurred during the 1970s. As described by Pinskey *et al.* (1974), 1975), astronauts on translunar trajectories or on long-duration space-station missions reported flashes of light visible during sleep periods when the spacecraft interior was darkened. It was realised that these flashes were the traces of high-energy cosmic rays – perhaps as massive as iron nuclei – upon the astronaut's optic nerves. In 1975 Eric Hannah estimated that several percent of an astronaut's neurons could be lost to nerve damage during a two-year round trip to Mars.

If cosmic-ray shielding were only a matter of attenuating a beam of non-reacting high-energy particles passing through the walls of a spacecraft hull, protecting a human crew from this hazard would be a comparitively simple task. All the designer would do would be to compare the intensity of an ionised particle beam (I_{beam}) after traversing a hull distance x_{hull} with its unattenuated intensity $I_{beam,0}$, using a form of the familiar exponential beam-attenuation equation (Segre, 1964):

$$I_{beam,x} = I_{beam,0} e^{-\mu_{beam} x_{hull}} \tag{11.1}$$

where μ_{beam} is the attenuation coefficient of the ion species in hull material.

Exercise 11.2 Consider an ion beam with an attenuation coefficient of $100\,\text{m}^{-1}$ traversing a spacecraft's hull. Plot the beam intensity relative to its unattenuated value as traversed hull thickness varies between 0 and 2 m. What happens when the beam-attenuation coefficient is halved or doubled?

Equation (11.1) is oversimplistic in practice (unfortunately) because high-energy ion beams interact with the matter they traverse. Secondary radiation produced by this interaction may actually be more injurious to biological tissues than the primary beam.

But a number of researchers have tackled this problem to estimate the spacecraft hull thickness (actually areal mass thickness) necessary to protect a human population from cosmic and solar ionising radiation. As part of a 1977 NASA design study of free-flying permanent space settlements (edited by Richard Johnson and Charles Holbrow), the hull areal mass thickness required to reduce maximum onboard radiation level to the terrestrial surface level of about $0.5\,\text{rem/year}$ (1 rem = 1 Reontgen Equivalent Man, a measure of radiation dosage to biological organisms) was estimated at $4,500\,\text{kg}\,\text{m}^2$. This corresponds to an aluminium hull thickness of 1.7 m.

However, this may be an overestimate of the amount of cosmic-ray shielding required to protect space explorers, because biological organisms have a varying tolerance to radiation before permanent damage – such as cancer – occurs. Since the current dosage limitations for long-duration astronauts is 50 rem per year, Rein Silberberg *et al.* argued in 1987 that considerably less shielding might be required. A 9-cm thick aluminium shield would reduce cosmic radiation dosage to about 35 rem per year – at least for spacecraft velcocities less than about 0.05 c.

Aluminium sheet may not be the best material for a cosmic-ray shield. According to Donald Radford *et al.* (1992), thorough knowledge of primary and secondary radiation interaction with matter may allow consideration of shields constructed using 'designer' composite materials, and a host of radiation-attenuating shapes imbedded in the shield structure such as spheres, fibres and platelets.

No matter what the wonders of advanced materials science achieve, however, interaction between the hull and interstellar gas at very high starship velocities would result in secondary neutron and gamma dosages that might be fatal. Passive cosmic-ray shielding by the spacecraft hull or structure becomes ineffective for high-velocity starships.

In the cold realm between the stars, superconducting cosmic-ray shields may see application for crew protection. As well as utilisation of the magsail, mini-magnetiosphere and ramscoops considered in previous chapters, mission designers might consider electromagnetic shields that have been designed for near-Earth space-colony application. One of these – the 'plasma core shield', – is discussed by Johnson and Holbrow.

An interstellar venture would probably make use of both passive and 'active' electromagnetic shielding. Payload might be strategically located to double as cosmic-ray shielding during such high-radiation manoeuvres as close stellar flybys.

11.2 OPTIONS FOR ONBOARD POWER BETWEEN THE STARS

Although most interstellar mission designers have understandably concentrated on problems of acceleration and deceleration, those who consider crewed missions must also devote some thought to the problems of onboard power during the long interstellar cruise. If onboard power (amounting to about 10 kW per crew-member, according to E. Bock *et al.*) is not provided, the crew will die as the ship's temperature approaches that of the near-absolute zero interstellar environment.

In one of the earliest attempts to design a crewed interstellar spacecraft, Gilfillan (referenced in Chapter 7) proposed an onboard fission reactor. Although it might at first seem that reliability problems during a centuries-long flight might doom such an approach, this is not necessarily the case. Natural nuclear reactors are produced under the Earth's surface by the random accumulation of radioactive isotopes, and such natural reactors can serve as the model for reliable fission reactors. Attempts to design reliable and safe fission reactors have been outlined by Freeman Dyson.

Since even a reliable, safe fission reactor will produce radioactivity, an interstellar crew might tether the reactor to the ship in a separate module. Gilfillan estimates that a fission reactor capable of supplying all power needs for a population of 25 during a three-century interstellar voyage might have a mass of 1,000 kg. From a 1999 analysis by James Powell *et al.*, high-reliability, low-mass fission power systems for space application are quite possible.

Another possibility, of course, is nuclear fusion or antimatter. If a spacecraft is propelled by one of these two nuclear options, a small fraction of the energy from the propulsion system reactor could be diverted to supply crew onboard power needs.

If the starship uses a magnetic ramscoop, a deceleration magsail, or a magnetic cosmic-ray deflector, there is another power option. As discussed by P. F. Smith in 1969, a superconducting solenoid or coil can serve as magnetic-energy storage battery as well as an ion deflection device.

As discussed in the section on breakthrough propulsion (Chapter 9), another possibility for onboard power is magnetic interaction between the ship and the interstellar medium. If the local interstellar-medium magnetic field lines are slow-moving relative to the starship, the ship could deploy a tether-like device to apply the induced EMF on charges in a conductor moving through a magnetic field. The work done on the moving charges could be converted to onboard power to supply the crew aboard a cruising starship, albeit at a small reduction in spacecraft kinetic energy relative to the local interstellar magnetic field.

Yet another possibility to supply the power needs of a starship's crew is a solar storage battery that could be charged whenever the craft is close to a star. One possibility for such a device is the 'light-sail windmill' proposed by Paul Birch in

1983. Such a device would consist of hyperthin and superstrong blades, with the blade aspect designed such that solar radiation pressure causes windmill spin-up during a close perihelion pass. (For further discussion of the light-sail windmill, see Matloff's (1985) paper on interstellar arks, the Matloff/Ubell (1985) paper cited in Chapter 4, Matloff's (1986) worldship paper, and Birch's (1985) correspondence in *Journal of the British Interplanetary Society*.)

Other onboard power alternatives will surely be suggested; but with all the available options, this aspect of interstellar mission planning will certainly not present us with a show-stopper.

11.3 CLOSED ENVIRONMENT LIFE SUPPORT SYSTEMS

One seeming paradox of space travel is the fact that as we move farther into the alien environments beyond the Earth, we must, for survival's sake, learn more and more about terrestrial ecosystems. In order to thrive beyond the Sun and between the stars, human pioneers will need to take a little of Earth with them.

Demonstration of the requirement for closed ecosystems is not difficult. In his 1969 consideration of human life in space, Mitchell Sharpe estimated that a human requires, each day, about 1 kg of oxygen, 4 kg of drinking water and 0.7 kg of food. If some environmental resource recycling is not carried out, a single human consumes 5.7 kg per day or about 2,000 kg per year of oxygen, water and nutrients. If a crew of ten participates in a century-duration interstellar venture, about 2×10^6 kg of these materials are required.

Various approaches to semiclosed ecosystems are described by Sharpe and many other authors, in which a combination of technology and biology is used to recycle oxygen and water, at the very least. But a major goal for interplanetary space colonisation and interstellar travel is the closed ecosystem, and much experimentation in this field remains to be performed.

Extensive experiments with nearly-closed ecological systems were performed in Russia during the 1970s, and have been described by Gitelson *et al.* (1976). As part of the BIOS 3 experiment, human volunteers lived for periods of half a year in a sealed, air-tight enclosure with dimensions $14 \times 9 \times 2.5$ m. The BIOS 3 complex was divided into four mini-ecospheres: two were occupied by higher plants, one by unicellular seaweed, and one by the 'crew'. The object was to biologically reprocess as much of the human-generated biological waste as possible, and regenerate oxygen, drinking water and food. Most of the energy required for the experiment was devoted to artificial lighting to trigger plant photosynthesis.

The atmosphere in BIOS 3 had a comparitively high concentration of CO_2, and the biomass gathered supplied the three-man 'crew' with 26% of their carbohydrate, 14% of their protein, and 2.3% of their fat requirements. The balance of the nutrient needs was supplied by vitamin supplements. Atmosphere was fully recycled, and about 95% of the crew's water needs. Extrapolating from the BIOS 3 results, the total non-regenerative life-support mass required for our ten-person, century-duration mission is reduced to about 100,000 kg.

To obtain a higher degree of ecosystem closure, perhaps approaching 100%, it is necessary to increase the number of crops, perhaps include some animals, and regulate the many environmental feedback loops. Jack Spirlock and colleagues review 1977 research in this field at NASA Ames Research Center as part of summer study dealing with space habitats.

Beginning in 1991, Biosphere 2 – a 12,750 m² enclosed ecological facility – was operated by an initially privately funded team in Oracle, Arizona. After a partially successful simulated two-year Mars flight with an 'astronaut' team living on board, Biosphere 2 was turned over to Columbia University, which has developed a programme to continue closed ecological system research (Wolfgang, 1995).

Research on space-qualified closed environmental modules, such as the one discussed by V. Blum *et al.* in 1995, has accelerated in recent years. It is certain that some of these ideas will be tried in near-Earth space during the era of the International Space Station. Within a decade or two we should know with a fair amount of confidence how to design a closed or nearly-closed interstellar ecosystem.

11.4 OF WORLDSHIPS AND INTERSTELLAR ARKS

Protection from dust-grain impacts and cosmic rays, and closed life-support systems, are significant aspects of interstellar-mission design, but they are certainly not sufficient in themselves. We must also consider the health of the human community onboard the interstellar craft, as the crew watches the Sun grow dimmer and dimmer astern of their lonely craft. How large a community of humans is necessary, and how big a ship is required to support them? Is it necessary to spin the starship to provide artificial gravity, or can an interstellar crew adapt to zero g and readapt to a planet's gravity field at journey's end?

The most spartan interstellar craft proposed was that of Gilfillan (cited in Chapter 7). With a mass of only 10^5 kg and a population of about 20, Gilfillan's starships would be non-rotating, zero-gravity space habitats. The final generation of a multi-century interstellar voyage would (hopefully) exercise rigorously to acclimatise themselves to the gravity of the destination planet before attempting a landing.

At the other extreme we have the space cities of Gerard K. O'Neill (1974 and 1977). These structures – which were proposed to house large human communities in deep space – could serve as the hub of an off-planet civilization supplying terrestrial energy needs using beamed solar power. Most of the enormous masses of the O'Neill colonies would be obtained from extraterrestrial resources – from the Moon, or as Brian O'Leary suggested in 1981, from near-Earth asteroids.

Figure 11.1 presents a representation of the O'Neill Model III Space Habitat, the smallest of his cylindrical space cities capable of shielding cosmic rays to terrestrial surface levels with the 10-cm aluminium walls, internal structure and internal atmosphere. Model III consists of two counter-rotating cylinders, each 10 km long and 2 km in diameter. Counter-rotation reduces precession so that the space city can remain oriented towards the Sun while providing its population with 1 Earth gravity on the habitat's inner rim.

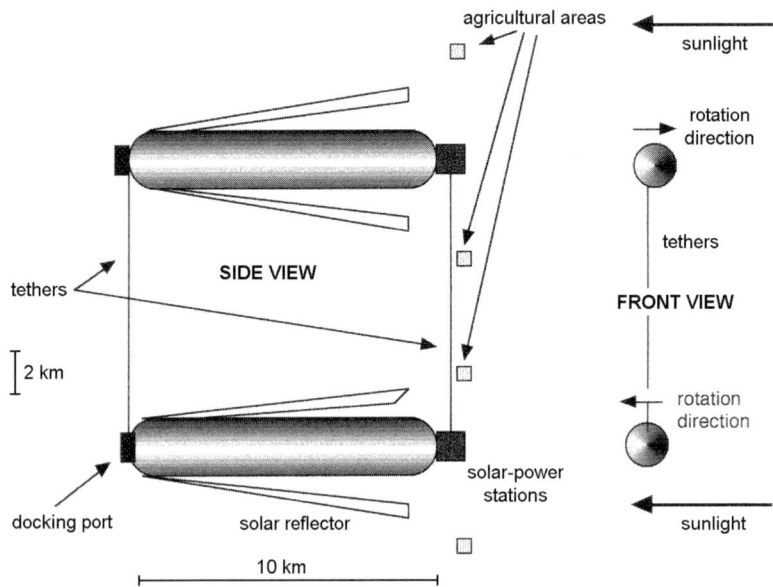

Fig. 11.1. The O'Neill Model III Space Habitat, which could be reconfigured as a worldship.

In its stationary mode, adjustable solar reflectors illuminate the interior of the habitat cylinders, as shown. The space cities could be configured in a low-population, exurban version in which agriculture was interior to the cylindrical habitats which have a population of about 100,000. A high-population, totally urbanised Model III could have a ring of agricultural modules external to the main habitat, as shown, and a population of around one million.

In both configurations, parklands, lakes and small forests could dot the inner rims of the habitats, adjacent to human population centres. Because fresh food would never be more than a few kilometres away, and climate could easily be adjusted, the need for refrigeration and house-heating/air-conditioning would be minimal. Inter-habitat transport could be by bicycle and electric vehicles because of the small distances involved. These factors should contribute to reducing pollution and increasing social stability.

To estimate rim centrifugal acceleration due to cylinder rotation, simply define acceleration experienced by a person on the habitat's inner rim (ACC_{hab}) as centripetal acceleration, apply the equation for angular velocity of the rotating habitat (ω_{hab}), and the definition of RPM = habitat revolutions per minute to obtain

$$ACC_{hab} = \omega_{hab}^2 R_{hab} = \left(\frac{2\pi}{60} RPM\right)^2 R_{hab} \text{ m s}^{-2} \qquad (11.2)$$

where R_{hab} is the habitat radius in metres. If our 1,000-m radius O'Neill Model III spins at about 1 revolution per minute, the population in the inner rim will experience Earth-normal gravity.

Exercise 11.3 Validate the derivation of equation (11.2); then plot a curve of artificial-gravity acceleration versus distance from the rim of our 1-km radius habitat cylinder.

The O'Neill Model III is large enough for weather patterns to be possible in the interior of the habitat. People might locate low-gravity recreation and industry near the centre, and choose to live on the colony rim. (For a recent treatment of artificial-gravity effects on humans, see the 1999 paper by Theodore W. Hall.)

As discussed by Johnson and Holbrow (1977), there are many possible alternatives to the cylindrical space habitat – including the sphere and torus. Elements of interior design and architecture of these structures should provide an interesting challenge for habitat designers, as suggested by Majorie Stuart in 1979.

When configured for interstellar travel, the solar reflectors would of course be closed, and non-solar onboard power would be required. Cylinder rotation might be suppressed during starship acceleration and deceleration.

It would also be possible to reduce the crew size, and therefore scale the space city to a smaller interstellar ark, as done by Matloff and Mallove in 1980. But an interstellar ark crew of 100 or so would not have the social resources of a larger worldship's crew in developing the destination solar system.

The sociological structure of a starship's crew presents interesting problems to the mission designer. While a small, fast ark could utilise a paramilitary organisation such as that found on contemporary voyages of exploration, this might not be the best social structure of a large worldship on a millennium-duration colonisation mission. Non-Western modes of social organisation might apply to such a venture, as suggested by those sociologists who contributed to *Interstellar Migration and the Human Experience*, a 1985 conference proceedings publication edited by Ben Finney and Eric Jones (cited in Chapter 1).

At the end of their long travels onboard their comfortable world ship, would the population be anxious to fly down to experience the uncertainties of life on the surface of an alien planet? Might they choose instead to enlarge their habitat within the non-threatening reaches of the new star's asteroid belts or comet clouds? Or might they – as suggested by M. G. deSan in 1981 – instead reprovision their worldship in the new solar system and immediately re-enter interstellar space in the role of perpetual nomads? Whatever the answers to these questions, there is one alternative to a thousand-year voyage onboard a world ship. Perhaps humans can sleep their way to the stars!

11.5 HIBERNATION FOR HUMANS: THE LONG SLEEP TO α CENTAURI

It would be ideal if humans were natural hibernators. But as reviewed by Mallove and Matloff in *The Starflight Handbook*, and by Maulin in *Prospects of Interstellar Travel* (both cited in Chapter 4), this is not the case. Also, cryogenic techniques that freeze a recently deceased organism to near-absolute zero are a good way to preserve dead meat, but methods of warming frozen brains back to room temperature without fatal brain-cell crystallisation do not exist. But if it could be achieved in humans, hibernation would slow human metabolism, reduce life-support requirements and possibly extend lifespan. As discussed by Sharpe (1969), hibernating mice can withstand both higher acceleration and radiation levels than can their alert colleagues.

Recently, Noriaki Kondo has discussed new research that may lead to a better understanding of hibernation. A hibernation-specific protein complex has been isolated in the chipmunk, a hibernating rodent. This material is produced in the chipmunk's liver and maintains the animal at constant body temperature during seasonal hibernation. Application of this knowledge to human hibernation is still to be demonstrated.

11.6 BIBLIOGRAPHY

Birch, P., 'Orbital Ring Systems and Jacob's Ladders-III', *Journal of the British Interplanetary Society*, **36**, 231–238 (1983).

Birch, P., 'Light Sail Windmills', (correspondence), *Journal of the British Interplanetary Society*, **38**, 527–528 (1985).

Blum, V., Andriske, M., Voeste, D., Behrens, H. and Keeuzberg, K., 'The C.E.B.A.S. Mini Module: Design of the Spaceflight Hardware, Scientific Experiments, and Future Aspects for Bioregenerative Life Support System Research', IAF/IAA-95-G.4.02.

Bock, E., Lambrou Jr., F. and Simon, M., 'Effect of Environmental Parameters on Habitat Structural Weight and Cost', in *Space Resources and Space Settlements, NASA SP-428*, NASA, Washington DC (1979), pp. 33–60.

deSan, M. G., 'The Ultimate Destiny of an Intelligent Species: Everlasting Nomadic Life in the Galaxy', *Journal of the British Interplanetary Society*, **34**, 219–237 (1981).

Dyson, F., *Disturbing the Universe*, Harper & Row, New York (1979).

Gitelson, I. I., Terskov, L. A., Kovrov, B. G., Sidko, Ya. F., Lisovsky, G. M., Okladnikov, Yu. N., Belyanin, V. N., Trubachov, L. N. and Rerberg, M. S., 'Life Support System with Autonomous Control Employing Plant Photosynthesis', *Acta Astronautica*, **3**, 633–650 (1976).

Hall, T. W., 'Artificial Gravity and the Architecture of Orbital Habitats', *Journal of the British Interplanetary Society*, **52**, 290–300 (1999).

Hannah, E. C., 'Meteoroid and Cosmic Ray Protection', in *Space Manufacturing Facilities (Space Colonies): Proceedings of the Princeton AIAA/NASA Conference*, May 7–9, 1975, ed. J. Grey, AIAA, Washington, DC (1977), pp. 151–158.

Johnson, R. D. and Holbrow, C. eds., *Space Settlements: A Design Study, NASA SP-413*, NASA, Washington DC (1977).

Kondo, N., 'Approach to Artificial Control of Hibernation', in *Missions to the Outer Solar System, 3rd IAA Symposium on Near-Term Advanced Scientific Space Missions*, ed. G. Genta, Aosta, Italy, 3–5 July 2000, Levrotto & Bella, Torino, Italy (2000), pp. 7–12.

Martin, A. R., 'Project Daedalus: Bombardment by Interstellar Material and its Effect on the Vehicle', supplement to *Journal of the British Interplanetary Society*, **31**, S116–S120 (1978).
Matloff, G. L., 'On the Potential Performance of Non-Nuclear Interstellar Arks', *Journal of the British Interplanetary Society*, **38**, 113–119, (1985).
Matloff, G. L., 'Faster Non-Nuclear World Ships', *Journal of the British Interplanetary Society*, **39**, 475–485 (1986).
Matloff, G. L. and Mallove, E. F., 'The First Interstellar Colonization Mission', *Journal of the British Interplanetary Society*, **33**, 84–88 (1980).
O'Leary, B., *The Fertile Stars*, Everest House, New York (1981).
O'Neill, G. K., 'The Colonization of Space', *Physics Today*, **27**, No.9, 32–40 (September 1974).
O'Neill, G. K., *The High Frontier*, Morrow, New York (1977).
Pinsky, L. S., Osborne, W. Z., Bailey, J. V., Benson, R. E. and Thompson, L. F., 'Light Flashes Observed by Astronauts on Apollo 11 Through 17', *Science*, **183**, 957–959 (1974).
Pinsky, L. S., Osborne, W. Z., Hoffman, R. A. and Bailey, J. V., 'Light Flashes Observed by Astronauts on Skylab 4', *Science*, **188**, 928–930 (1975).
Powell, J. R., Maise, G., Paniagua, J., Ludewig, H. and Todosow, M., 'Compact, Ultra Lightweight Nuclear Thermal Propulsion Engines for Planetary Science Missions', presented at NASA/JPL/MSFC/AIAA Annual Tenth Advanced Space Propulsion Workshop, Huntsville, AL, April 5–8, 1999.
Radford, D. W., Sadeh, W. Z. and Cheng, B. C., 'Composite Materials Microstructure for Radiation Shielding', IAF-92-0324.
Segre, E., *Nuclei and Particles*, Benjamin, New York (1964).
Sharpe, M. R., *Living in Space*, Doubleday, Garden City, New York (1969).
Silberberg, R., Tsao, C. H., Adams Jr., J. H. and Letaw, J. B., 'Radiation Hazards in Space', *Aerospace America*, **25**, No.10, 38–41 (1987).
Smith, P. F., 'The Technology of Large Magnets', in *A Guide to Superconductivity*, ed. D. Fishlock, Macdonald, London UK (1969), Chapter 4.
Spurlock, J., Cooper, W., Deal, P., Harlan, A., Karel, M., Modell, M., Moe, P., Phillips, J., Putnam, D., Quattrone, P., Raper Jr., C. D., Swan, E., Taub, F., Thomas, J., Wilson, C. and Zeitman, B., 'Research Planning Criteria for Regenerative Life-Support Systems Applicable to Space Habitats', in *Space Resources and Space Settlements, NASA SP-428*, NASA, Washington DC (1979), pp. 13–30.
Stuart, M. L., 'Aesthetic Considerations in Bernal Sphere Design', AIAA 79-1428, in *Proceedings of Fourth Princeton/AIAA Conference on Space Manufacturing Facilities*, Princeton, NJ, May 14–17, 1979, AIAA, Washington DC (1979).
Wolfgang, L., 'Biosphere 2 Turned Over to Columbia', *Science*, **270**, 1111 (1995).

12

Conscious spacecraft

And now, out among the stars, evolution was driving toward new goals. The first explorers of Earth had long since come to the limits of flesh and blood; as soon as their machines were better than their bodies, it was time to move. First their brains, and then their thoughts alone, they transferred into shining new homes of metal and plastic. In these, they roamed among the stars. They no longer build spaceships. They were spaceships.

Arthur C. Clarke, *2001: A Space Odyssey* (1968)

As the previous chapter reveals, sending humans to the stars will not be easy, barring some unexpected breakthrough. Perhaps humans could sleep their way to the stars, waking up like some Rip Van Winkle after a sleep of centuries or millennia. Or perhaps they could travel for a few centuries in a crowded interstellar ark or for a millennium in a more spacious, slower and expensive worldship.

Might there be another way? Might we accomplish all of humanity's interstellar goals without sending (physical) humans at all? Many authors have speculated upon this subject. Some of the early work is reviewed in Mauldin's *Prospects for Interstellar Travel* (cited in Chapter 4).

If our goals in space travel are to explore and gather information about strange, new worlds, there are alternatives to crewed starships. We might instead send a fleet of tiny, self-reproducible intelligent robots with programmed instructions to 'phone home'. Instead, if we intend to colonise nearby star systems, there might be simpler alternatives to the world ship. We might choose to send a nanotechnological fleet with a few carefully frozen fertilised eggs of humans and other organisms. Upon arrival at the new solar system, the nanobots could respond to preprogramming and construct a 'robot nanny' to raise the first generation of human, animal and plant colonists.

Finally, what if we pursue interstellar travel as a means of furthering the realm of consciousness, as suggested by Olaf Stapledon is his 1937 classic science fiction *The Star Maker*. Perhaps, like the protagonist of this novel, we might travel to the stars as non-corporeal beings, perhaps as 'virtual humans' downloaded into tiny supercomputers (as suggested by Frank Tipler in 1994).

12.1 THE VON NEUMANN MACHINE: CAN THE COMPUTER EAT THE GALAXY?

In 1966, mathematician John von Neumann suggested that it might someday become possible to construct tiny, self-reproducing automata. These small 'nanobots' would be analogous to organic creatures. They might be launched to the stars aboard microscopic, slow starcraft. Upon arrival, their internal programming could enable them to rendezvous with each other, construct larger robots, explore the new environments and transmit their findings home. It is not inconceivable that their programming could function like the DNA of biological organisms. They could then construct a new generation and fly off to new interstellar targets.

Von Neumann machines have been featured in many science-fiction novels. In Arthur C. Clarke's *2001: A Space Odyssey* and its sequels, the monoliths that direct terrestrial and Europan evolution, and eventually reproduce to convert Jupiter into a star, are von Neumann machines.

Constructing such a micro- or nano-sized intelligent, self-organising and self-reproducing entity may be more difficult in practice than on paper or celluloid. As speculated by Salvatore Santoli in 1995, the structure of a nanobot might be more like that of a biological entity than a technological tool.

As discussed by P. A. Hansson in 1992, computer chips functioning in a manner analogous to the human brain – neural chips – are a real possibility. And as suggested by Hansson in 1995, intelligent, interacting automata constructed using these neural chips might even satisfy our criteria as self-aware, self-conscious entities.

Such speculations do not necessarily belong to the farthest future. Computer technology has been evolving at breakneck speed, and the concept of supercomputers constructed at the quantum level is not an impossibility. If such tiny intellects can be constructed and made radiation-resistant, humans could easily pepper our outer Solar System and nearby stellar systems with inexpensive, solar-sail launched nanobots within a millennium.

We have found no clear evidence of alien von Neumann machines within our Solar System. Because of the apparent ease of 'occupying' an entire galaxy using this technology, a number of authors, including Barrow and Tipler (1986), have argued that humanity may be the first technological species to emerge in our Galaxy.

Exercise 12.1 Consider a class of von Neumann machines launched by a species with the intent to occupy the Galaxy. Assume that the nanobots can cross between neighbouring star systems at an average separation of 5 light years in 900 years, and a 'stop-over' of 100 years is required in each newly-reached star system to construct the next generation of star-faring nanobots. How long will it take the descendents of the original nanobot generation to travel 100,000 light years and cross the Galaxy?

Perhaps the answer to the apparent absence of von Neumann machines in our Solar System is to be found more in the realm of ethics than in technological possiblities. What would motivate an intelligent race to build machines to 'conquer' the Galaxy?

Another point is that of control. Even if we can construct and launch the first generation of machines designed to colonise the Galaxy, might the programme 'mutate' in further generations? In other words, might the von Neumann machines or their descendents 'rebel' against their original purpose and commence to 'do their own thing'?

12.2 THE CRYOGENIC STAR-CHILD

Instead of occupying the entire Galaxy, an ethically-advanced interstellar species might instead desire to spread its organic and robotic offspring through only a tiny fraction of space. Michael Mautner (1996) has suggested a possible near-term driver for the development of cryogenic biotechnology. Human institutions seem to be inherently unstable and human numbers currently threaten the survival of numerous plant and animal species. Perhaps these can be preserved by storing genetic materials at temperatures close to absolute zero, in the depths of space.

If such a technology proves to be feasible and is developed, the long-term storage in space of human genetic material – perhaps fertilised eggs – is certainly a possibility. Then a marriage of von Neumann and cryogenic techniques might be accomplished. Nanoprobes to the nearest stars might contain 'seed' packets for humans and other terrestrial organisms (Hansson, 1996). Upon arrival, the nanobots would be programmed to use the space resources of the new solar system to construct nurseries in which to raise the first generation of 'organic colonists'.

The ethics of colonising nearby stellar systems with people, animals and plants raised by robots instead of humans is certainly open to debate. If such a programme is accomplished, might the trauma experienced by trading the security of the robotic nursery for the unknown hazards of a newly discovered planet's surface prompt the emergence of a 'Garden of Eden' myth?

12.3 THE VIRTUAL STAR TRAVELLER

Perhaps the most exotic possibility presented by the nanocomputer is the following. We cannot claim to possess a science of consciousness, although many authors such as Roger Penrose (1989) have attempted to address the 'mind–body problem'. One way to consider the mind, according to Tipler (1994), is as a form of software designed to work on an organic computer – the brain.

If we can treat the mind–brain as an elaborate type of information-processing unit, it might be possible to ultimately create non-organic equivalents to the brain – devices of the same or greater complexity as the human brain that could directly interface with human and animal brains. Assuming that interface and hardware problems can be solved, and that humans and machines can be linked at the neuronal level, we are faced with an interesting possibility. Perhaps the software defining the human mind (or soul?) might be downloaded to the computer.

Even from the perspective of this book, this is a very far-out idea. It addresses issues of philosophy, theology and psychology as well as the more mundane concerns of the computer and spaceship designer.

Tipler (1997) has suggest that a 100-g nanocomputer payload may be capable of containing the memories of as many as 10^4 human brains. It is strange to imagine that within such a tiny structure, the essences of a host of human personalities might interact as if in a small town, in a computer-generated virtual reality.

Even though human technology is a long way from creating such virtual space travellers and their environments, the idea is indeed compelling. Whether or not Tipler is correct in his opinion that all human star travellers will be virtual rather than physical, the ethics and appropriateness of the concept should be widely debated.

12.4 BIBLIOGRAPHY

Barrow, J. D. and Tipler, F. J., *The Anthropic Cosmological Principle*, Clarendon Press, Oxford, and Oxford University Press, New York (1986).

Hansson, P. A., 'Conscious Satellites as an Exploratory Tool', IAA-92-0232.

Hansson, P. A., 'Exobiology, SETI, von Neumann, and Geometric Phase Control', *Journal of the British Interplanetary Society*, **48**, 479–483 (1995).

Hansson, A., 'Towards Living Spacecraft', *Journal of the British Interplanetary Society*, **49**, 387–390 (1996).

Mautner, M., 'Space-Based Genetic Conservation of Endangered Species', *Journal of the British Interplanetary Society*, **49**, 319–320 (1996).

Penrose, R., *The Emperor's New Mind*, Oxford Uuniversity Press, New York (1989).

Santoli, S., 'Nanostructured Undecidable Semantic Microrobots for Advanced Extrasolar Missions', IAA-95-IAA.4.1.04.

Tipler, F. J., *The Physics of Immortality*, Doubleday, New York (1994).

Tipler, F. J., 'Ultrarelativistic Rockets and the Ultimate Future of the Universe', in *NASA Breakthrough Propulsion Physics Workshop Proceedings, NASA/CP-1999-208694*, ed. M. Millis, NASA Lewis (Glenn) Research Center, August 12–14, 1997.

von Neumann, J., *Theory of Self-Reproducing Automata*, University of Illinois Press, Chicago, IL (1966).

13

Meeting ET

Civilizations hundreds or thousands or millions of years beyond us should have sciences and technologies so far beyond our present capabilities as to be indistinguishable from magic. It's not that what they do can violate the laws of physics; it's that we will not understand how they are able to use the laws of physics to do what they do.

Carl Sagan, *The Cosmic Connection* (1973)

Like Marshall Savage in 1992, we can design elaborate programmes aimed at the establishment of small, high-tech, self-sufficient human communities, first on Earth, then in interplanetary space and finally throughout the Galaxy. But all such plans hinge upon a major uncertainty: what happens if we establish direct, physical contact with an extraterrestrial (ET) civilization?

If we look to terrestrial history for guidance, we can only become depressed. When technological equals meet in the same geographical setting – as did the Hebrew refugees from Bronze Age Egypt and the Philistines from the Minoan–Mycenean world about 3,000 years ago – centuries of warfare are the probable result. And when a technologically advanced terrestrial civilization (such as Western Europeans) contacts a less advanced society (such as the American Indians), the less advanced people are at a severe disadvantage.

Astronomers who participate in the search for extraterrestrial intelligence (SETI) have long pondered the ethics of ET contact. Engineers planning star voyages would be well advised to follow the same course. Most of the speculation regarding diplomatic and sociological consequences of ET contact, such as the relevant sections in books by Lemarchand and McDonaugh, have considered the consequences of radio contact. But what if contact comes through starships (ours or ET's) rather than radio?

In a situation of that type, the expertise of sociologists and other humanists might be of more use than that of the physical scientists. For sociological perspectives regarding ET contact, the reader is urged to consult the work of Albert Harrison, Peter Schenkel and Donald Tarter. Douglas Vakoch of the SETI Institute believes that human–ET contact might be facilitated through the use of interstellar messages communicating spiritual as well as scientific principles.

13.1 ARE STARSHIPS DETECTABLE?

Until humanity begins to construct its own large interstellar spacecraft, astronomers might investigate suspect objects remotely, to ascertain whether ET is operating near our Solar System. Although energetic antimatter starships are the most easily observed, even small probes can be detected under certain conditions.

Detecting energetic starships

No attempt will be made here to discuss observable properties of breakthrough-physics starships propelled by spacetime warps or zero-point energy. This is not because such ships are impossible; it is simply that we do not yet know enough about them to characterise their observational parameters.

We begin our examination of starship detection therefore, with energetic 'conventional' starships – those accelerated by nuclear processes (fission, fusion and antimatter) and decelerated by magsails. Robert Zubrin (1996) has considered the spectral signature of such craft. As Zubrin describes, the signature of an energetic starship will be different from an astrophysical object because the starship's position and speed will be inconstant. Future astronomers using very large terrestrial and space telescopes might detect such objects in many wavebands across the electromagnetic spectrum.

The problem is the number of photons per second that will fall on a terrestrial detector, since energetic starships will probably be spherically symmetric radiators. Zubrin predicts that a starship emitting 10^{18} W of 200 MeV gamma rays at a distance of 1 light year from the Sun will result in a flux of about 7.5 photons per m^2 per year on a gamma-ray detector near the Earth, which renders detection difficult, if not impossible.

> *Exercise 13.1* Using the approximate conversions 1 MeV $\cong 1.6 \times 10^{-13}$ J, and 1 light year $\cong 10^{16}$ m, confirm Zubrin's result discussed in the preceding paragraph. Assume that the gamma-ray-emitting starship is a spherically-symmetric radiator.

The gamma-ray emissions from an antimatter starship might be undetectable, but since visible light will also be radiated from the ship's hot exhaust, and since the visible-light radiation might be better collimated than the gamma flux, such an exhaust stream might be telescopically visible over light years.

Fusion rockets will emit mostly in the X-ray range, and this flux could be detected over distances of a few light years if the orbiting X-ray detector is large enough and sensitive enough. Although a fission-electric drive also emits X-rays, some of the radiated electromagnetic flux is in the visible/infrared spectral ranges.

If our hypothetical energetic ET starship decelerates using a magsail, the interaction between the magsail electromagnetic field and the interstellar plasma will result in some electromagnetic emission. Zubrin estimates that such electromagnetic emissions will be mostly in the X-ray and radio spectral ranges. If the decelerating starship is large and fast enough, magsail-caused radio-frequency emissions might be detectable over a range of many light years by a 6-km orbiting (or lunar) radio detector.

Detecting laser/maser light sails

If ET travels prefers non-nuclear interstellar travel, he might utilise a laser or maser light sail. If the starship is near enough and the laser/maser is powerful enough, reflections from the sail might be observable as a fast-moving and accelerating monochromatic 'star'. However, detection will depend on sail shape and orientation as well as other physical factors.

Therefore, it is not as easy to model the spectral signature of these craft as it is energetic nuclear craft. A starship accelerated using lasers or masers may be easier to detect during deceleration if a magsail is used.

Detecting non-energetic starships

If ET chooses to cross the abyss in slow worldships accelerated and decelerated by solar sails, he could be detected as he departs his home solar system. Gregory Matloff and John Pazmino have considered the detectability of such craft over interstellar distances.

Since solar sails for worldship acceleration and deceleration will probably be of planetary dimensions, migrating worldships will appear to terrestrial observers as planet-like objects that reflect the light of their parent stars, but moving in a direction radially outward from the parent star. At least when they are closest to their parent stars (and therefore brightest), such objects will be seen to accelerate.

Detecting such objects will be possible when the planned generation of 'terrestrial-planet-observing' telescopes comes on line. If we can detect a terrestrial planet close to its parent star, reflected light from a worldship sail should also be visible. Of course, such detections would be serendipitous, since an accelerating worldship will not long remain within the confines of its home solar system. Rather than concentrating on random stars within the solar neighbourhood for signs of an occasional expedition, it might be better to examine the vicinity of stars from which an entire civilization could be emigrating.

If ET has migrated from his home system, and if some of his representatives are nearby, might they be detected even if they choose not to communicate? Michael Papagiannis has speculated that we might detect stationary worldships within our Solar System, if such exist. Assuming that such craft house organic spacefarers, their interior temperature must be elevated to support life. Such craft would stand out from life-sized asteroids or comet nuclei because of their excess infrared radiation.

13.2 MOTIVATIONS OF STAR-TRAVELLING EXTRATERRESTRIALS

In order to successfully search for ET starships, we must consider ET's motivations as well as his detectable technology. If some form of spacetime warp is feasible and very inexpensive, technologically advanced extraterrestrials might travel very far from home. Such voyages may be of commercial, research or even recreational nature. If travel by such means is developed by extraterrestrials, and if alien civilizations are close enough to our Solar System, an extraterrestrial presence on Earth might be commonplace.

But travel by such means requires a new physics as well as a new technology. If we confine instead ourselves to physically possible propulsion systems such as the antimatter rocket and the fusion ramjet, a society that solves the technological and economic problems of moderately fast star-travel might send out occasional voyages of exploration.

In their classic treatment of intelligent cosmic life, I. S. Shklovskii and Carl Sagan speculate that a stable, long-lived spacefaring civilization might time such voyages in order to bring some novelty to the home civilization and perhaps prevent stagnation. One interstellar mission per century might be attempted by such an advanced culture. Perhaps the best articles of trade with such a culture might be art, philosophy and music.

A typical non-spacefaring civilization might expect to be visited at intervals of about a millennium. Shklovskii and Sagan suggest that we might re-examine some of the myths of pre-scientific humanity in case these are based upon occasional ET visitations.

Even a stay-at-home culture with no interest in communicating with its interstellar neighbours might seriously consider interstellar travel when its home star begins to leave the main sequence. As suggested by Matloff in a web essay (www.accessnv/nids) for the National Society for Discovery Science (NIDS), and later amplified in a 1999 IAA paper co-authored by Matloff, Schenkel and Marchan, the nearest candidate solar-type star expanding from the main sequence is β Hydri, at a distance of about 21 light years.

Stars expand and increase in luminosity as they leave the main sequence, so they become better 'launch pads' for solar-sail-propelled worldships as they begin to destroy life on their inner planets. Basing their estimate upon the frequency of random close stellar approaches, and the assumption that an emigrating civilization would seek to reduce interstellar travel time by directing their worldships towards the nearest suitable stars, Matloff et al. (1999) estimate that the outer reaches of our Solar System may have been colonised if one out of every 10,000 stars in the sky has a long-lived spacefaring civilization. If we detect a radio-silent cluster of Oort Cloud or Kuiper Belt objects with a tell-tale infrared excess, should we attempt communication? And if we do, who speaks for Earth?

It is also quite possible that there is an alien robotic presence in the Solar System, even if starships have not visited us or if worldships have not taken up residence. Perhaps silent probes are monitoring our development and beaming data home, as suggested by Alan Tough (referenced in Chapter 5) and by Richard Burke-Ward

(2000). If a smart, long-lived ET observation robot is discovered within our Solar System, there will of course be efforts to communicate with it. Again, should such communication be attempted, and who are to speak for us?

We are clearly at the beginning of an era when humanity's robotic reach will begin to extend towards the stars; and as scientists and engineers, diplomats, poets and artists, we must become involved in this endeavour.

13.3 BIBLIOGRAPHY

Burke-Ward, R., 'Possible Evidence of Extra-Terrestrial Technology in the Solar System', *Journal of the British Interplanetary Society*, **53**, 2–12 (2000).

Harrison, A. A., *After Contact*, Plenum, New York (1997).

Lemarchand, G. A., *El Llamado de las Estrellas* (in Spanish), Lugas Editorial S.A., Buenos Aires, Argentina (1992).

Matloff, G. L. and Pazmino, J., 'Detecting Interstellar Migrations', in *Astronomical and Biochemical Origins and the Search for Life in the Universe*, ed. C. B. Cosmovici, S. Bowyer and D. Werthimer, Editrice Compositori, Bologna, Italy (1997), pp. 757–759.

Matloff, G. L., Schenkel, P. and Marchan, J., 'Direct Contact with Extraterrestrials: Possibilities and Implications', IAA-99-IAA.8.2.02.

McDonaugh, T., *The Search for Extraterrestrial Intelligence*, Wiley, New York (1987).

Papagiannis, M. D., 'An Infrared Search in the Solar System as Part of a More Flexible Search Strategy', in *The Search for Extraterrestrial Life: Recent Developments*, ed. M. D. Papagiannis, D. Reidel, Boston, MA (1985), pp. 505–511.

Savage, M. T., *The Millennial Project*, Empyrean, Denver, CO (1992).

Schenkel, P., 'The nature of ETI, its Longevity and Likely Interest in Mankind: The Human Analogy Re-Examined', *Journal of the British Interplanetary Society*, **52**, 13–18 (1999).

Shklovskii, I. S. and Sagan, C., *Intelligent Life in the Universe*, Holden–Day, New York (1966).

Tarter, D. E., 'Alternative Models for Detecting Very Advanced Extra-Terrestrial Civilizations', *Journal of the British Interplanetary Society*, **49**, 291–296 (1996).

Vakoch, D. A., 'Communicating Scientifically Formulated Spiritual Principles in Interstellar Messages', IAA-99-IAA.9.1.10.

Zubrin, R., 'Detection of Extraterrestrial Civilizations via the Spectral Signature of Advanced Interstellar Spacecraft', *Journal of the British Interplanetary Society*, **49**, 297–302 (1999).

Afterword

When deep-space exploration was underfunded, a book like this would remain current for years; but this is no longer the case. As a NASA Marshall Space Flight Center (MSFC) faculty fellow, I have attended many recent conferences in this field, and am therefore able to include here the most recent developments.

Discovered extrasolar planets now number almost 50, the most exciting of which is the reported jovian companion to ϵ Eridani.

Geoffrey Landis, of Ohio Aerospace Institute, critiqued use of superconductor shielding to create unidirectional currents during the July 2000 Joint Propulsion Conference in Huntsville, Alabama. He believes that such a shield around a wire will shift the force produced by an external magnetic field from wire to superconductor, but the entire system will still behave like a closed current loop. Tethers and electrostatic techniques may be better for the applications discussed here.

During the July 2000 IAA meeting in Aosta, Italy, Roger Lenard of Sandia Labs presented a NEP-propelled Kuiper Belt Object sample return mission. The Italian scientist Salvatore Santoli argued that KBOs could instead be explored using nano-probes. I will propose a non-nuclear KBO mission during STAIF 2001.

Also during the Aosta 2000 IAA meeting, fine artists became involved in the design of message plaques to be carried onboard future interstellar probes. In 'Message from Earth' – an exhibition curated by my wife C Bangs – more than thirty visual artists displayed their ideas. Some of this work was later presented at MSFC. The enthusiasm for deep-space exploration seems to be growing rapidly, even among non-scientists. Scientists including Robert Forward and Les Johnson are also considering how to store vast quantities of visual information in a message plaque.

Gregory L. Matloff
25 August 2000

Nomenclature

Variables

ABS_{sail}	light-sail fractional absorption
ACC_{hab}	artificial gravity on the interior rim of a rotating space habitat
ACC_{ion}	acceleration of interstellar ion
$ACC_{laser\ sail}$	acceleration, laser sail
$ACC_{rp,sail}$	radiation pressure acceleration of a lightsail
A_{sc}	solar-cell area
A_{scoop}	ramscoop field area
a_{mesh}	one half of perforated sail wire mesh width
BA	Bond albedo
$B_{ep,sw}$; $B_{ep,lw}$	short- and longwave approximations for planet/star brightness ratio
B_{ism}	interstellar magnetic field intensity
B_{torus}	magnetic field of toroidal ramscoop
D_{ion}	distance travelled within toroidal ramscoop by interstellar ion
$D_{las-ship,max}$	maximum separation, laser and starship
$D_{planet-star}$	separation of a planet from its primary star
$E_{nep,beam}$	NEP ion beam energy, in electron volts
F	force
F_{ion}	force on interstellar ion
$F_{loop,mag}$	magnetic force on current loop
f_{fl}	Fresnel lens focal length
f_{qr}	(energy/particle from a ZPE machine)/(energy/particle from a fusion reactor)
f_{rec}	flux received from celestial object
f_u	fraction of solar sail unfurled at perihelion
H_{ex}	exospheric scale height
$I_{beam,0}$, $I_{beam,x}$	ion beam intensity before and after traversing hull thickness
I_{mag}	magsail supercurrent

Nomenclature

I_{sp}	specific impulse
i_{loop}	loop current
I_{wire}	wire current in toroidal ramscoop
$J_{nep,beam}$	NEP ion beam current, in amps
$J_{pm,mag}$	magsail current to mass density ratio
K_{rair}	interstellar/onboard fuel consumption for ram-augmented interstellar rocket
KE_{nf}	nuclear fuel kinetic energy
$KE_{s,grf}$	kinetic energy of a relativistic starship, according to Earthbound observer
L_{loop}	loop length
$L_{s,grf}$ and $L_{s,srf}$	length measured in unaccelerated and accelerated reference frames
M_{cable}	cable mass
M_{cloud}	interstellar cloud mass
M_{coll}	collapsar mass
M_{con}	atmospheric gas molecular mass
M_{dust}	mass of an interstellar dust grain
M_f	fuel mass
$M_{f,am}$	antimatter fuel mass
M_{if}	ion fuel mass
M_{ion}	mass of interstellar ion
M_j	Jupiter's mass
M_{mag}	magsail mass
$M_{nep,lp}$	NEP low-power power-conditioning subsystem mass
$M_{nep,rad}$	NEP radiator mass
M_{nf}	nuclear fuel mass
M_0	unfuelled spacecraft mass
M_{obj}	mass of celestial object
M_{part}	mass of particle beam
M_{probe}	probe mass
M_s	ship mass
$M_{s,grf}$ and $M_{s,srf}$	M_s measured in unaccelerated and accelerated reference frames
M_{th}	thruster mass
m_{bol}	bolometric magnitude
MR	mass ratio of a Newtonian rocket
MR_{rel}	mass ratio of a relativistic rocket
NEP	nuclear-electric propulsion
N_{wire}	number of turns in toroidal ramscoop wire
NZ_{fl}	number of zones in a Fresnel lens
OT_{ir}	planet atmospheric infrared optical thickness
P_{laser}	laser power
P_{nep}	NEP fission reactor power, in kilowatts
$P_{planet,absorbed}$	radiant power absorbed by planet
P_{sp}	specific power

$P_{s,grf}$	momentum of a relativistic starship, according to Earthbound observer
P_{th}	thruster power
q_{ion}	interstellar ion charge
Q_{nep}	nuclear-fuel/nuclear-inert-fuel consumption ratio
Q_{net}	net electrical charge
R_{array}	solar-collecting array radius
R_{au}	distance from Sun, in Astronomical Units
R_{cent}	distance to centre of gravitating body
$R_{e,au}$	distance from Earth, in Astronomical Units
R_{estr}	electrostatic turning radius
R_{ex}	planet exospheric radius
R_{fl}	Fresnel lens radius
R_{hab}	radius of a cylindrical space habitat
R_{init}, R_{fin}	initial and final positions
$R_{init,au}$	initial distance from Sun, in Astronomical Units
$R_{\lambda,mesh}$	mesh reflectance at wavelength λ
R_{mag}	magsail physical radius
R_{mtr}	magnetic turning radius
R_{peri}	perihelion distance
R_{planet}	planet radius
R_{sail}	sail radius
$R_{sl,au}$	Sun-laser separation, in Astronomical Units
R_{sch}	the Schwartzchild radius of a collapsar
R_{star}	star radius
R_{Sun}	solar radius
R_{tele}	telescope radius
$R_{trans,sail}$	fractional reflectance of transmissive light sail
RAD	celestial object radius
REF	celestial object reflectivity
REF_{sail}	light-sail reflectivity
REM	Roentgen Equivalent Man (or Mammal): biological radiation-effect measure
RPM	rotation rate of a rotating space habitat, in revolutions per minute
RP_{sail}	radiation pressure on a light sail
r_{torus}	distance from centre of toroidal ramscoop
$S_{fl,out}$	spacing, Fresnel lens outer zone
T_{eff}	effective black-body temperature
$T_{eff,planet}$	effective planet black-body temperature
T_{ex}	exospheric temperature
$T_{\lambda,mesh}$	perforated light-sail mesh fractional transmission
Th_{nep}	NEP ion-engine thrust, in Newtons
TE	total orbital energy
$T_{trans,sail}$	fractional transmittance of transmissive light sail
t_{star}	star surface temperature

Nomenclature

$T_{surface}$	planet surface temperature
$T(k)$	absolute temperature of sail material
$(TS)_{cable}$	cable tensile strength
t	time
t_{burn}	NEP nuclear engine operating time, in years
t_{ion}	time for ion to be deflected from r_{torus} to toroidal ramscoop centre
t_{mesh}	mesh thickness
$t_{s,grf}$ and $t_{s,srf}$	time measured in unaccelerated and accelerated reference frames
U_{rair}	parameter used in ram-augmented interstellar rocket kinematics consideration
V_1, V_2, V_3	velocity prior to, at and after periapsis
V_{beam}	particle beam velocity
$V_{con,av}$	mean thermal molecular velocity of an atmospheric gas
V_{cr}	cruise velocity
V_e	exhaust velocity
$V_{e,if}$	exhaust velocity, ion fuel
$V_{e,nf}$	exhaust velocity, nuclear fuel
V_{fp}	ramjet runway fuel-pellet velocity
V_{in}, V_{fin}	initial and final velocities
V_{mag}	ship velocity during magsail deceleration
V_{para}	escape of parabolic velocity
$V_{para,peri}$	perihelion parabolic velocity
$V_{peri,obj}$	parabolic velocity at periapsis of celestial object
V_{prs}	probe velocity relative to Sun
V_s	ship velocity
V_{srm}	ship velocity relative to local interstellar magnetic field
$V_{srm,per}$	ship velocity component perpendicular to local interstellar magnetic field
W	weight
WHR	waste heat radiated
$W_{planet,emit}$	radiant flux emitted by planet
W_{tele}	reflected light energy/second entering telescope
x_{hull}	thickness of spacecraft hull traversed by ionised cosmic radiation beam
$\beta_{e,rock}$	rocket exhaust velocity/c
β_{fp}	V_{fp}/c
β_{in}, β_{fin}	$V_{in}/c, V_{fin}/c$
β_s	V_s/c
γ	$[1 - \beta_s^2]^{-1/2}$
ΔKE	change in kinetic energy
$\Delta \theta$	angular separation
ΔV_{peri}	change in velocity at perihelion
δ_{skin}	skin depth
ε_{ei}	electrical to nuclear-inert exhaust energy conversion efficiency

ε_{ia}	efficiency of converting electrical energy to ion exhaust kinetic energy
ε_{laser}	laser efficiency
ε_{lrc}	laser-to-electricity conversion efficiency
ε_{ne}	nuclear-electric energy conversion efficiency
ε_{nf}	nuclear energy fraction transferred to nuclear exhaust
ε_{sc}	solar cell efficiency
$\varepsilon_{trans,sail}$	emissivity of transmissive light sail
Φ_{nf}	mass/energy conversion efficiency
η_{sail}	sail lightness factor
λ	light wavelength
λ_{laser}	laser wavelength
μ_{beam}	attenuation coefficient of an ion beam in matter
μ_{mesh}	perforated light-sail mesh cross-sectional circumference
μ_{mtp}	momentum transfer efficiency
θ	angle subtended by distant celestial object
ρ_{cable}	cable density
ρ_{dust}	interstellar dust grain density (grains per cubic metre)
ρ_{ion}	interstellar ion density
ρ_{in}	interstellar ion mass density
$\rho_{in,grf}$	interstellar ion mass density measured in an unaccelerated reference frame
σ_{eff}	sailcraft areal mass thickness
ω_{cloud}	angular velocity of interstellar cloud
ω_{hab}	angular velocity of a rotating space habitat (in radians per second)
ψ	trajectory bend angle

Constants

c	speed of light, 3×10^8 m s^{-1}
g	gravitational acceleration at Earth's surface, 9.8 m s^{-2}
G	Universal gravitational constant, 6.67×10^{-11} N m^2 kg^{-2}
h	Planck constant, 6.63×10^{-34} J s
k	Boltzmann constant, 1.38×10^{-23} J kg^{-1}
M_{Sun}	solar mass, 1.99×10^{30} kg
μ_{fs}	permeability of free space, 1.26×10^{-6} Henry metre (or Newtons amps^{-2})
q_{elect}	electron charge, -1.6×10^{-19} c
S_c	solar constant, 1,368 W m^{-2}
σ	Stefan–Boltzmann constant, 5.67×10^{-8} W m^{-2} K^{-4}

Interplanetary/interstellar distance units

Astronomical Unit (AU)	1.5×10^{11} m
Light year	9.46×10^{12} km
Parsec	3.26 light years

Glossary

Aerobraking A method of spacecraft deceleration by momentum-transfer collision with molecules in the outer fringe of a planet's atmosphere.
Aneutronic reaction A thermonuclear fusion reaction in which few or no neutrons are produced.
Antimatter Charge-reversed particles with the same mass as their corresponding 'ordinary' matter counterparts.
Astrometry The branch of astronomy dealing with stellar motions and positions.
Black body An object that absorbs all electromagnetic radiation that strikes it.
Brown dwarf An object intermediate in size between a giant planet and tiny star.
Burn fraction The fraction of a nuclear reaction's output energy that can be transferred to the kinetic energy of a spacecraft exhaust.
Collapsar A gravitationally collapsed celestial object.
Cosmic ray A highly energetic charged subatomic particle with an extraterrestrial origin.
Doppler shift The shift in light wavelength or colour caused by the radial motion of the light source.
Ecosphere The region in a planetary system in which life on an Earth-like planet might form. At the inner and outer ecosphere boundaries, the planet's oceans boil or freeze.
Emissivity A measure of how closely an object approximates a black body, which has an emissivity of 1.
Exhaust velocity The velocity of a spacecraft motor's reaction products as measured from the spacecraft.
Exosphere The upper limit of a planet's atmosphere; its boundary with space.
Fission A nuclear reaction in which a massive atomic nucleus splits.
Fresnel lens A thin-film optical component made up of alternating plastic layers and capable of redirecting a beam of light in a lens-like manner.
Fusion A nuclear reaction in which low-mass atomic nuclei are merged.
Gravity assist A manoeuvre in which a spacecraft increases orbital energy by interacting with the gravitational field of a celestial object.

176 Glossary

Gravity lens Optical focusing and amplification by a gravitational field.

Heliosphere The realm of the Sun's influence on interplanetary particles and fields, bounded by the heliopause at about 100 AU from the Sun.

Insolation Received solar radiation per unit area.

Interferometer A telescope in which several or many small mirror components share a common focus and thereby simulate resolving power of an equally large, single-mirror telescope.

Interstellar ark A spacecraft requiring many generations to complete an interstellar journey.

Kuiper Belt A region of icy, sub-planetary objects between about 30 and 50 AU from the Sun.

Laser/maser sail A spacecraft propelled by the pressure of a collimated beam of radiation from a laser or a microwave maser.

Magsail A spacecraft decelerator operating by the magnetic reflection of ions.

Mass ratio The quotient of a spacecraft's fully fuelled to unfuelled mass.

Muon A short-lived sub-nuclear particle intermediate in mass between an electron and a proton.

Nanobot A robot constructed at the molecular size level.

Near-Earth objects (NEOs) A population of small (~kilometer-diameter), Earth-approaching (or impacting) objects of asteroidal and cometary origin.

Neutrino An uncharged sub-nuclear particle, with little or no mass, that is unreactive with conventional matter.

Nuclear-electric propulsion (NEP) A spacecraft propulsion technique in which nuclear energy is used to ionise and electrically accelerate exhaust fuel.

Nuclear-pulse propulsion Spacecraft propulsion using exploded nuclear 'devices' or fusion micropellets as exhaust fuel.

Nuclear-thermal propulsion A spacecraft propulsion technique in which nuclear energy is used to heat and accelerate a spacecraft's exhaust fuel.

Oort Cloud A spherical region extending to perhaps 100,000 AU from the Sun in which as many as one trillion comets may exist.

Optical thickness A measure of the opacity of a planet's atmosphere to light.

Orbital energy For an object in open or closed orbit around a larger celestial object, the sum of kinetic and potential energies (relative to the central object).

Panspermia The theory that life originated at one or a few locations in the Universe and then spread to other celestial habitats either naturally or artificially.

Parabolic velocity Escape velocity.

Particle-beam propulsion Spacecraft propulsion by momentum transfer from an impinging beam of accelerated particles.

Perforated light sail A solar or laser sail reduced in mass by the incorporation of many tiny perforations with dimensions less than a wavelength of light.

Photometry A branch of astronomy dealing with star brightness.

Pion A short-lived sub-nuclear particle intermediate in mass between an electron and a proton.

Proper motion Motion of a celestial object across the celestial sphere.

Quantum ramjet An interstellar ramjet obtaining energy from the universal vacuum.

Radial motion The motion of celestial object towards or away from the observer.
Radiation pressure Electromagnetic photons carry momentum, although they have zero mass.
Ram-augmented interstellar rocket (RAIR) A spacecraft obtaining reaction mass from the interstellar medium and energy from onboard nuclear sources.
Ramjet A spacecraft obtaining both reaction mass and energy from the interstellar medium.
Ramscoop A device used to collect fuel or reaction mass from the interstellar medium.
Rayleigh's criterion A measure of resolving or collimating power of an optical device.
Resolving power (resolution) The measure of a telescope's ability to observe objects of small angular size.
Rocket A spacecraft propelled by the exhaust of energised onboard fuel.
Schwartzchild radius The distance from the centre of a collapsar at which escape velocity exceeds the speed of light.
Solar constant Power received per square metre from the Sun by an object in space at 1 AU from the Sun and oriented normal to the sunlight.
Solar-electric propulsion A spacecraft propulsion technique in which solar energy is used to ionise and electrically accelerate exhaust fuel.
Solar sail A thin-film spacecraft propelled by the momentum of impinging photons of sunlight.
Solar-thermal propulsion Energy of sunlight is used to heat and accelerate a spacecraft's fuel.
Solar wind An interplanetary stream of ions flowing outward from the Sun.
Specific impulse A measure of rocket performance.
Terraforming The modification of a planet's environment to make it more Earth-like.
Tether A long cable or wire unfurled in space.
Thrustless turn A manoeuvre in which spacecraft interaction with an interstellar magnetic field effects a change in direction.
Worldship A spacecraft requiring many generations to complete an interstellar journey with a simulated Earth-like environment.
ZPE Zero-point energy obtained from the universal vacuum.
ZPE laser A laser pumped by ZPE.

Index

14 Herculis, 140
36 Ophiuchi, 142
47 Ursae Majoris, 139
51 Pegasi, 139
55 Cancri, 139
70 Ophiuchi, 142
70 Virginis, 140
82 Eridani, 142

α Centauri, 16, 21, 26, 33, 50, 52, 57, 62, 71, 73, 79, 86, 97, 110, 141, 142, 153
adaptive optics, 137
Advanced Space Propulsion, 78
aeolipile, 36
albedo, 17
Alcubierre, Miguel, 127
Aldrin, Buzz, 21, 52
Alenia Spazio, 27
Allegheny Observatory, 134
Alvarez, Luis, 3
Alvarez, Walter, 3
Ames Research Center, 21, 60, 61, 151
Anderson, Carl, 76
Anderson, Poul, 101
Andrews, Dana, 88, 98
aneutronic fusion, 76
angular momentum, 14
antigravity, 122
antimatter, 26, 76, 83, 149
antiproton, 76, 77
Apollo 12, 63
Archytas, 35

asteroid assembler, 62
asteroid belt, 2
asteroids, 6, 18, 21, 35, 62, 151
Aston, Graeme, 70
Astrochicken, 59, 62
astrometry, 133
ASTROsail, 27
atmosphere, 18
atmospheric optical depth, 18
Aurora project, 29

β Hydri, 108, 142, 164
β Pictoris, 134
bacteria, 16, 63
Bakes, Emma, 21
ballistic missiles, 5
Banfi, V., 126
Barnard's Star, 79, 134
Barnes, John, 52
beam-core engine, 78
beamed-energy sailing, 83
Bekey, Ivan, 120
Belbruno, Edward, 6, 46, 97
Berry, Adrian, 127
beryllium, 52
BIOS 3, 150
Biosphere 2, 151
Birch, Paul, 149
black hole, 29, 121, 126
black-body (effective) temperature, 16, 17
Bloomer, John, 97
Blum, V., 151

Bock, E., 149
bolometric magnitude, 8
Bond albedo, 17
Bond, Alan, 74, 104
Bond, R.A., 71
boron, 107
Boss, Alan, 135
brain, 159
Brenner, D., 62
British Interplanetary Society, 74
brown dwarf, 135, 139
Brusa, E., 30
Burke-Ward, Richard, 164
Burnham, Darren, 58
Bussard, Robert, 101, 103

Campbell, Bruce, 134
carbon, 104
carbon microtruss, 30
carbon-12, 104
Casimir effect, 122
Casimir, H.B.G., 122
Cassanova, Robert, 121
Cassenti, Brice, 50, 77, 112, 113, 114
Cassini, 40, 65
Chapline, G., 77
Charon, 21
Chinese rocket, 36
circumsolar zone, 25, 26
Clarke, Arthur C., 158
closed ecosystems, 150
collapsar, 126, 127
colliding beam, 76
Columbia University, 151
comets, 2, 16, 18, 22, 25, 31, 35, 62
 Halley, 46
 Shoemaker–Levy 9, 5, 22
comparative planetology, 20
Congreve, William, 36
cosmic rays, 26, 63, 147
Cosmovici, Cristiano, 31
crew health, 147
Croswell, Ken, 134
cryogenics, 159

δ Pavonis, 142
Daedalus, 72, 74
Davies, Paul, 63

Davis, E., 127
Debye–Hückel screening, 91, 113
Deep Space 1, 40
deSan, M.G., 153
deuterium, 74, 76
Dick, S.J., 134
Dirac, Paul, 76
directed panspermia, 9
Dole, Stephen, 142
Dorian civilization, 1
Drexler, Eric, 58
Druyan, Ann, 6
Dyson, Freeman, 59, 73, 149

ε Eridani, 134, 142, 167
η Cassiopeiae, 142
Earth, 15, 17, 18, 19, 20, 21
Earth-defence projects, 31
Eberlein, Claudia, 123
ecosphere, 16, 141
effective (black-body) temperature, 16, 17
Ehricke, Kraft, 25, 44
Eichorn, Heinrich, 134
Einstein Cross, 28
Einstein Ring, 28
electric-field meters, 26
electrodynamic tether, 120
electromagnetic shielding, 148
electromagnetic waves, 27
electron, 77
Electron Power Systems Inc, 121
Energy Science Laboratories Inc, 30
Eshelman, V., 27
ethics, 9
Europa, 16, 20, 142
evolution, 3
exosphere, 19
extrasolar planets, 13, 29, 133, 167
extrasolar space, 25
extraterrestrial life, 31, 161

Fearn, David, 70
Fennelly, Alphonsis J., 87, 92, 103, 112
Fermilab, 78
'fire' zone, 16
Fisher, D.E., 134
Fisher, M.J., 134
fission, 66, 83, 112, 149

Index

Fleischmann, Martin, 118
flux, 8, 16
focal probes, 29
Fogg, Martyn, 143
Fomalhaut, 134
Forward, Robert, 77, 90, 94, 96, 122, 167
fossil record, 3
Fresnel lens, 94
Friedman, Louis, 46
Froning, David, 122, 123, 125
Ftaclas, Christ, 138
fuel pellets, 110
Fuller, Buckminster, 62
Fullerene nanotube, 62
fusion, 26, 66, 83, 111, 149
fusion plasma, 75

G186, 139
Gaidos, G., 78, 79
galaxies, 29
Galileo, 16, 40
gamma rays, 68, 76, 77, 162
Garner, Charles, 30
gas-dynamic mirror fusion, 75
Gatewood, George, 134, 138
general relativity, 126
Geneva Observatory, 134
Genta, G., 30
Genz, Henning, 122
Gilfillan, Edward, 86, 149, 151
Glanz, James, 123
Glenn Research Center, 118, 127
Gliese 229, 135
Gliese 876, 140
globular clusters, 29
Globus, Al, 61
Goddard, Robert, 36
Goldin, Dan, 30, 79
Goldsmith, Donald, 138
gravity assist, 26, 29, 40, 43
gravity focus, 27
gravity lens, 27
Greenhouse Effect, 18
gunpowder, 36

Hahn, Joseph, 31
Haisch, Bernard, 123
Halyard, R.J., 79

Hannah, Eric, 147
Hansson, Anders, 60, 62
Harrison, Albert, 161
HD 10697, 140
HD 37124, 140
HD 75289, 139
HD 114762, 140
HD 134987, 140
HD 168443, 140
HD 177830, 140
HD 187123, 139
HD 192263, 139
HD 209458, 135
HD 209458, 139
HD 210277, 140
HD 217107, 140
HD 222582, 139
Heidmann, J., 29
heliopause, 22, 26, 121
heliosphere, 22, 25
helium-3, 74
helium-4, 104
Hero of Alexandria, 36
hibernation, 153
Hiten, 46
Holbrow, Charles, 148, 153
Howe, S.D., 78
HR 810, 140
HR 7703, 142
HR 8832, 142
Hubble Space Telescope, 8, 135, 136
hydrogen, 74
hydrogen-fusing interstellar ramjet, 101

'ice' zone, 16
inertial electrostatic confinement, 75
infrared interferometer, 138
infrared space telescope, 134
interferometer, 9, 135
International Academy of Astronautics, 29
International Space Station, 151
interstellar ark, 86, 151, 153
interstellar clouds, 16
interstellar colonisation, 2
interstellar dust, 145
interstellar ramjet, 68, 97, 101
interstellar rocket, 97
ion (solar-electric) propulsion, 26, 35, 39
ion fuel, 69

Ionian civilization, 1
iridium, 3
Italian Energy Board, 118

Jackson, A.A., 97, 108, 110, 112
Jaffe, L.D., 26
Jet Propulsion Laboratory, 26, 46
Johnson, Les, 30, 119, 167
Johnson, Richard, 148, 153
Jones, Eric, 96
Jupiter, 2, 5, 6, 7, 16, 17, 18, 19, 20, 21, 22, 40, 41, 43, 44, 74, 121, 142
Jupiter gravity assist, 26, 29

Kammash, T., 76
Kare, Jordin, 97
Knowles, Timothy, 30
Koczor, Ron, 122
Kondo, Noriaki, 154
Kuiper Belt, 21, 25, 31, 71, 164, 167
Kuiper Belt Explorer, 31

Lalande 21185, 138
Lamoreaux, Steven, 123
Landis, Geoffrey, 127, 167
Langton, N.H., 112
laser ramjet, 108
laser-electric propulsion, 26
laser/maser sails, 26, 84
lasers, 61
Lawes, R.A., 60
laws of physics, 13
Lee, M.-J., 76
Lee, Russell, 119
Leifer, Stephanie, 30
Lenard, Roger, 32, 112, 167
Lenz's law, 119
Lesh, J.R., 61
Lewis, R.A., 78
Lick Observatory, 134
life, 140
life support systems, 150
lifezone, 20
light-sail windmill, 149
Lipinski, Ronald, 32, 70, 112
Lippincott, Sarah, 134
lithium, 107

long-baseline astrometry, 26
lunar soil, 74

Maccone, Claudio, 27, 29, 45, 127
Maclay, Jordan, 123
macro spacecraft, 58
magnetic field, 120, 121
magnetic surfing, 118
magnetometer, 26
magnetosphere, 25
magnitude, 8
magsail, 83, 87
Manhattan Project, 66
Marcy, Geoff, 135
Mariner 10, 21, 40
Mars, 2, 6, 15, 16, 17, 18, 19, 20, 21, 35, 58, 142
Mars Global Surveyor, 58
Mars Pathfinder, 2, 58
Marsden, Brian, 6
Marshall Space Flight Center, 31, 75, 118, 119, 120, 167
Martin, Anthony R., 71, 112, 146, 147
masers, 61
mass extinction, 3
Massier, P.F., 77
Matloff, Gregory L., 4, 6, 29, 50, 52, 62, 84, 87, 92, 93, 97, 103, 108, 110, 112, 123, 153, 163
matter/antimatter annihilation, 66, 76
Mauldin, John, 147
Mautner, Michael, 9, 159
Mayor, Michel, 134
McCoy, Bruce, 119
McKendrie, T., 62
McNutt, R.L., 33
Medusa, 72, 75
Mehta, Alkesh, 119
Mercury, 15, 17, 18, 21, 29, 40, 122
meteorites, 3
micro spacecraft, 58
microorganisms, 9
micropellets, 110
microphotography, 60
Mileikovsky, Curt, 57
Miley, George, 128
Millis, Marc, 127
Minami, Yoshinari, 127
mini-magnetosphere, 121

mining, 2
Mir, 46
Mocci, Gabriele, 30
Moeckel, W.E., 128
Moody, Fred, 127
Moon, 2, 35, 36, 151
Morgan, D.L., 77
Morgan, J.A., 128
Mourou, G.A., 127
muon, 77

nano-payloads, 9
nanobiology, 60
nanobots, 62, 158
nanocables, 62
nanoprobes, 167
nanoribbons, 62
nanostarships, 63
nanotechnology, 58, 59
nanotubes, 62
NASA, 26, 46, 72, 75, 79, 118, 119, 120, 121, 136, 148, 151, 167
   Breakthrough Propulsion Office, 127
   heliopause sail, 30
   Institute for Advance Concepts, 121
   interstellar initiative, 30
   Interstellar Probe, 30
   Kuiper Belt Explorer, 31
National Society for Discovery Science, 164
near-Earth objects (NEO), 2, 4, 6, 7
Neptune, 7, 17, 21, 31
neural chips, 158
neutrino, 77
neutron star, 121
neutrons, 68
New York City Technical College, 119
Next Generation Space Telescope, 138
nitrogen, 104
nitrogen-13, 104
Noble, Robert J., 71
Nock, K., 40
Noever, Robert, 122
Nordley, Gerald, 98, 110
Norem, P.C., 90, 96
nuclear power, 65
nuclear-electric propulsion, 26, 32, 69
nuclear-pulse propulsion, 26, 72

O'Leary, Brian, 151
O'Neill Model III Space Habitat, 151, 152
O'Neill, Gerard K., 2, 135, 151
Oberth, Herman, 36
Observatoire de Paris, 138
occulting disk, 135, 136
Ohio Aerospace Institute, 167
onboard power, 149
Oort Cloud, 2, 6, 22, 25, 26, 32, 71, 83, 164
optical cameras, 26
optical depth, 18
Orion, 72
Oro, Juan, 31
oxygen, 104

panspermia, 9
Papagiannis, Michael, 163
Parks, Kelly, 6
particle-beam propulsion, 97
Pazmino, John, 163
Pecchioli, M., 89
pellets, 110
Penrose, Roger, 159
perforated light sail, 92
Perry, M., 127
photon drive, 68
Piantelli, Francesco, 118
pion, 77
Pioneer 10, 22, 25, 35
Pioneer 11, 22, 25, 35, 40
Pioneer 12, 40
planet effective temperature, 16
planet habitability, 16
planetary atmosphere, 18
planetary rovers, 58
Planetary Society, 4
planetesimals, 15
planetology, 20
plasma core shield, 148
Pluto, 21, 26, 35
Polynesians, 1
Pons, Stanley, 118
positron (antielectron), 76, 77
Powell, Conley, 107
Powell, James, 149
primeval nebula, 13
proton, 77, 104, 107
protoplanet, 15

protostar, 15
Puthoff, Harold, 123

quantum ramjet, 123
Quasat, 27

ρ CrB, 139
Radford, Donald, 148
radiant flux, 16
radio astronomy, 26
radio transmitters, 61
radioactive isotopes, 72
radioactive-isotope thermal generators, 29
radioisotope-electric propulsion, 71
radiometers, 26
ram-augmented interstellar rocket, 68, 76, 104
ramjet, 68, 97, 101, 108, 110, 123
ramjet runway, 110
ramscoop, 101, 112
Rayleigh's criterion, 8
Remo, John, 4
Robosloth, 97
robotic probes, 2, 9, 26, 52
rocket, 35, 97, 111
Roman, Nancy, 136
Rostoker, Norman, 76
rovers, 58

σ Draconis, 142
Sagan, Carl, 6, 103, 164
Salmon, Andy, 58
San Francisco State University, 135
Sandia Labs, 167
Sandquist, Eric, 140
Sanger, Eugene, 76
Santoli, Salvatore, 30, 53, 63, 158, 167
Saturn, 7, 17, 20, 21, 40, 43, 44
Savage, Marshall, 161
Scaglione, S., 30, 53
Schenkel, Peter, 161
Schilling, Govert, 135
Schneider, Jean, 138
Segre, Emilio, 76
SETI (search for extraterrestrial intelligence), 161
SETI Institute, 161
SETIsail, 27

Seward, Clint, 121
Sharpe, Mitchell, 150
Shlovskii, I.S., 164
Shmatov, M.I., 112
Silberberg, Rein, 148
Singer, Clifford, 97
Smith, G.A., 78
Smith, P.F., 149
Smithsonian Air and Space Museum, 73
solar cells, 29
solar constant, 7, 16
solar gravisphere, 26
solar gravity assist, 43
solar magnetosphere (heliosphere), 25
solar radiant flux, 16
solar sails, 6, 9, 25, 26, 35, 46, 83, 92, 121
Solar System, 2, 13, 15, 21
solar wind, 22, 74
solar-electric (ion) propulsion, 26, 35, 39
solar-system formation, 13
solar-thermal drive, 25
Solem, Johndale, 75
sonoluminescence, 123
Space Shuttle, 26, 37, 38
spacecraft kinematics, 124
spacetime warps, 126
special relativity, 124
special-relativistic star drive, 122
speckle interferometry, 137
spectral classification, 140
spectrometer, 26
Spirlock, Jack, 151
Sproul Observatory, 134
STAIF 2000 conference, 50
STAIF 2001 conference, 167
Stapledon, Olaf, 157
star formation, 13
starship detection, 162
Starwisp, 96
Stefan–Boltzmann law, 17
Stuart, M.L., 153
superconducters, 120, 167
supernova, 13
supernova remnants, 134
Surveyor, 63
survival, 3
Swarthmore University, 134
Szames, Alexandre, 122

τ Boö, 139
τ Ceti, 110, 142
Tarter, Donald, 161
Tarter, Jill, 136
TAU mission, 26
telescopes, 7
terraforming, 142
terrestrial life, 31
Terrile, Richard, 138
thermal generators, 29
thrustless turning, 90
Tipler, Frank J., 158, 159, 160
toroidal ramscoop, 112
Tough, Alan, 61, 63, 164
tritium, 74, 76
Tsiolkovsky, Konstantin, 36
Tunguska event, 4

υ Andromedae, 139
Ulysses, 40
United States Air Force, 72
University of British Columbia, 134
University of California at Berkeley, 3
University of Illinois, 123, 128
University of Siena, 118
University of Southern Florida, 134
University of Washington, 121
Uranus, 7, 17, 21

V2, 36
Vakoch, Douglas, 161
Van Cleve, Jeff, 138
van de Kamp, Peter, 134
Vega, 134
Venus, 15, 17, 18, 19, 20, 21, 35, 40, 58

von Neumann machine, 158
von Neumann, John, 158
Voyager 1, 2, 22, 25, 32, 35, 40, 43, 45, 52
Voyager 2, 2, 22, 25, 35, 40, 44, 45, 52
Vulpetti, Giovanni, 29, 50, 89

Wan Hu, 36
Ward, William, 31
'water' zone, 16
Wells, H.G., 36
white dwarf, 141
white hole, 127
Whitmire, D.P., 97, 108, 110, 112
Wilford, John Nobel, 135
Williams, I.P., 6
Winglee, Robert, 121
Winterberg, F., 73, 74
Wisdom, J., 6
'wooden dove', 35
worldship, 145, 151
worm hole, 127

X-rays, 68, 162

Yenne, Bill, 58
Yilmaz, Huseyin, 122
Yucatan, 3

Zadnik, Mario, 63
zero-point energy, 122
Zito, R.R., 77
Znamia 2, 46
Zubrin, Robert, 87, 88, 162